ANNALS OF THE NEW YORK ACADEMY OF SCIENCES

Volume 927

EDITORIAL STAFF

Executive Editor
BARBARA M. GOLDMAN

Managing Editor
JUSTINE CULLINAN

Associate Editor
STEPHANIE J. BLUDAU

The New York Academy of Sciences
2 East 63rd Street
New York, New York 10021

THE NEW YORK ACADEMY OF SCIENCES
(Founded in 1817)

BOARD OF GOVERNORS, September 2000 – September 2001

BILL GREEN, *Chairman of the Board*
TORSTEN WIESEL, *Vice Chairman of the Board*
RODNEY W. NICHOLS, *President and CEO* [ex officio]

Honorary Life Governors
WILLIAM T. GOLDEN   JOSHUA LEDERBERG

JOHN T. MORGAN, *Treasurer*

Governors

ELEANOR BAUM	D. ALLAN BROMLEY	KAREN BURKE
	LAWRENCE B. BUTTENWIESER PRAVEEN CHAUDHARI	
JOHN H. GIBBONS	MICHAEL GOLDEN	RONALD L. GRAHAM
ROBERT G. LAHITA	JACQUELINE LEO	WILLIAM J. McDONOUGH
JOHN F. NIBLACK	SANDRA PANEM	RICHARD RAVITCH
RICHARD A. RIFKIND	SARA LEE SCHUPF	JAMES H. SIMONS

HELENE L. KAPLAN, *Counsel* [ex officio]   NANCY B. EISENBERG, *Interim Secretary* [ex officio]

THE ONSET OF NONLINEARITY IN COSMOLOGY

ANNALS OF THE NEW YORK ACADEMY OF SCIENCES
Volume 927

THE ONSET OF NONLINEARITY IN COSMOLOGY

Edited by James N. Fry, J. Robert Buchler, and Henry Kandrup

The New York Academy of Sciences
New York, New York
2001

Copyright © 2001 by the New York Academy of Sciences. All rights reserved. Under the provisions of the United States Copyright Act of 1976, individual readers of the Annals are permitted to make fair use of the material in them for teaching or research. Permission is granted to quote from the Annals provided that the customary acknowledgment is made of the source. Material in the Annals may be republished only by permission of the Academy. Address inquiries to the Permissions Department (editorial@nyas.org) at the New York Academy of Sciences.

Copying fees: For each copy of an article made beyond the free copying permitted under Section 107 or 108 of the 1976 Copyright Act, a fee should be paid through the Copyright Clearance Center, Inc., 222 Rosewood Drive, Danvers, MA 01923 (www.copyright.com).

∞ The paper used in this publication meets the minimum requirements of the American National Standard for Information Sciences—Permanence of Paper for Printed Library Materials, ANSI Z39.48-1984.

Library of Congress Cataloging-in-Publication Data

The onset of nonlinearity and cosmology / edited by James N. Fry, J. Robert Buchler, and Henry Kandrup.
    p. cm. — (Annals of the New York Academy of Sciences, ISSN 0077-8923 ; v. 927)
    Includes bibliographical references and index.
    ISBN 1-57331-324-6 (cloth : alk. paper) — ISBN 1-57331-325-4 (pbk. : alk. paper)
    1. Cosmology—Congresses. 2. Nonlinear theories—Congresses. I. Fry, James N. II. Buchler, J. R. (J. Robert) III. Kandrup, Henry E. IV. Florida Workshops in Nonlinear Astronomy and Physics (15th : 2000 : Gainesville, Fla.) V. Series.

Q11.N5 vol. 927
[QP980]
500 s—dc21
[523.1]
                                                                                     2001031491

GYAT / PCP
Printed in the United States of America
ISBN 1-57331-324-6 (cloth)
ISBN 1-57331-325-4 (paper)
ISSN 0077-8923

ANNALS OF THE NEW YORK ACADEMY OF SCIENCES
Volume 927
June 2001

THE ONSET OF NONLINEARITY IN COSMOLOGY

Editors and Conference Organizers
JAMES N. FRY, J. ROBERT BUCHLER, AND HENRY KANDRUP

[This volume comprises the proceedings of a conference entitled the **15th Florida Workshop in Nonlinear Astronomy and Physics: The Onset of Nonlinearity in Cosmology**, held during February 17–19, 2000, in Gainesville, Florida.]

CONTENTS

Preface	vii
A Random Walk through Models of Nonlinear Clustering. *By* RAVI K. SHETH	1
A New Angle on Gravitational Clustering. *By* ROMÁN SCOCCIMARRO	13
The Transition to Nonlinearity and New Constraints on Biasing. *By* ROMAN JUSZKIEWICZ AND ENRIQUE GAZTAÑAGA	24
Peculiar Velocity Surveys: Optimal Moments Analysis. *By* HUME A. FELDMAN, RICHARD WATKINS, ADRIAN L. MELOTT, AND PATRICK GORMAN	43
Non-gaussianity versus Nonlinearity of Cosmological Perturbations. *By* LICIA VERDE	54
The Cosmological Mass Function in the Zel'dovich Approximation. *By* SERGEI F. SHANDARIN	70
Lensing of the CMB: Non-Gaussian Aspects. *By* MATIAS ZALDARRIAGA	84
Cosmic Statistics of Statistics: N-point Correlations. *By* ISTVÁN SZAPUDI	94
Dark Matter Caustics. *By* P. SIKIVIE AND W. KINNEY	102
Nonlinear Gravitational Growth Inside and Outside the Standard Cosmology. *By* E. GAZTAÑAGA AND J.A. LOBO	110

An Attempt to Do without Dark Matter. *By* WILLIAM H. KINNEY AND
MARTINA BRISUDOVA 127

Emergence of Anomalous Distributions in Disordered Systems.
By K.A. MUTTALIB AND P. WÖLFLE 136

The Onset of Nonlinearity in Cosmological Structure. By J.N. FRY AND
CHUNG-PEI MA 143

INDEX OF CONTRIBUTORS 159

Financial assistance was received from:

- DEPARTMENTS OF PHYSICS AND ASTRONOMY,
  THE COLLEGE OF LIBERAL ARTS AND SCIENCES, AND
  THE OFFICE OF RESEARCH AND GRADUATE PROGRAMS
  AT THE UNIVERSITY OF FLORIDA

> The New York Academy of Sciences believes it has a responsibility to provide an open forum for discussion of scientific questions. The positions taken by the participants in the reported conferences are their own and not necessarily those of the Academy. The Academy has no intent to influence legislation by providing such forums.

Preface

On the largest observable scales, the universe appears to be nearly homogeneous and isotropic, with small fluctuations that are easily and accurately treated in linear perturbation theory. On small scales, local variations are large, and the behavior of the distribution of matter is studied by numerical simulation. In the intermediate regime, neither suffices completely: nonlinearity in the evolution can no longer be ignored, but finite volume and resolution effects make simulation difficult as well. This volume focuses on these intermediate scales, the onset of nonlinearity, statistical techniques for identifying its effects (anomalous non-Gaussian distributions, higher order correlations), and analytic methods (mode coupling, phenomenological approximations) for its study. Such considerations may be of interest on other similar settings, such as the conductance fluctuations in disordered condensed matter systems or the onset of hydrodynamical turbulence.

The 15th Florida Workshop in Nonlinear Astronomy and Physics took place 17–19 February 2000 in Gainesville, Florida, bringing together researchers from around the world working in the area of nonlinear cosmological structure and other areas of physics, with the hope that different perspectives on common problems would lead to an exchange of methods and techniques to the advantage of both. Such cross-disciplinary exchanges are becoming increasingly common and productive.

Support for the speakers and workshop infrastructure was provided by the Departments of Physics and Astronomy, the College of Liberal Arts and Sciences, and the Office of Research and Graduate Programs at the University of Florida.

We are additionally grateful to the members of the Editorial Department of the New York Academy of Sciences, and particularly to Stephanie Bludau for her gracious and professional help in guiding the book through the press.

— THE EDITORS

A Random Walk through Models of Nonlinear Clustering

RAVI K. SHETH

NASA/Fermilab Astrophysics Group, MS 209, Batavia, Illinois 60510-0500, USA

ABSTRACT: A few simple models of the mass function of collapsed objects are presented. The emphasis is on apparently unrelated models which end up giving the same answer for the number density and merger histories of virialized clumps. Models of the spatial distribution of the clumps and how they can be used to model the spatial distribution of the mass are briefly discussed.

KEYWORDS: Clusters; Cosmology; Dark matter; Theory

1. INTRODUCTION

In what follows I describe a few simple, toy models of the growth of clustering. The emphasis will be on the interrelations between these models, more than on the exact agreement with the results of simulations. All the models I discuss are hierarchical in the sense that small things form first and big things form later by mergers of the small things; there is no fragmentation.

2. THE MASS FUNCTION AND MERGER HISTORIES OF COLLAPSED OBJECTS

In the following discussion I focus on how one might estimate the probability that a randomly chosen particle belongs to a clump that contains m particles. All the models use the initial spatial distribution of the mass to estimate how clusters grow. This means that the models are most likely to be accurate if the mass was cold initially.

2.1. Gravity as an Effective Length-Scale

To begin, consider a distribution of particles arranged at random along a line—we will consider three-dimensional distributions shortly. This random distribution is supposed to represent the initial spatial distribution. We wish to model what gravitational evolution does to this distribution. Presumably, gravity being a force of attraction, near neighbors will begin to move toward each other. Suppose that if two particles collide, they merge. We would like to estimate the number of clumps con-

---

Address for correspondence: Ravi Sheth, Fermilab MS209, P.O. Box 500, Batavia, IL 60510. Voice: 630/840-3665.
sheth@fnal.gov

taining m particles at a time t after the initial distribution of single particles started to cluster. There are two natural choices for the order in which things happened.

The first is to assume that clusters grow by a process similar to percolation. The intuitive idea is that gravity can rearrange things only on small scales at early times, but on increasingly larger scales at later times—the length scale associated with gravity increases with time. To model this, draw a line outward from each particle; assume that the length of this line increases monotonically with time. Two particles are said to be friends and to have merged with each other if the line from one of them touches the other. If we define clumps by requiring that friends-of-friends are in the same clump, then a set of clusters is defined as a function of link-length. The size of a typical cluster must grow as the friend-of-friend length increases; large link-lengths mean late times, reflecting the fact that massive clusters are not present initially, but are more common later.

If we keep track of the clumps at time t_1 which are part of a larger clump at $t_2 > t_1$, then we have stored some information about the merger history of an object. This means that we can start to address questions such as: How does the distribution of clump sizes evolve? Are the most massive objects at t_2 made of the most massive objects that were present at t_1? How different are the merger histories of objects that contain the same number of particles?

The link-length model above provides analytic answers to all these questions if the initial distribution of points was Poisson. In the one-dimensional case, the probability that a clump contains m particles when the critical link-length is l equals

$$\eta(m, l) = \left[\prod_{i=1}^{m-1} \int_0^l \exp(-\bar{n}x_i) n dx_i \right] e^{-\bar{n}l} = e^{-\bar{n}l} [1 - e^{-\bar{n}l}]^{m-1} \quad \text{where } m \geq 1, \quad (1)$$

and $\bar{n}$ denotes the average density of particles. The term in square brackets can be understood as follows. The probability that the particles are at $(dl_1, \ldots, dl_n)$ equals $\Pi \bar{n}\, dl_i$ times the probability that there are no particles in between: $\Pi \exp(-\bar{n}\, l_i)$, where l_i denotes the distance between particles i and $i + 1$. If we change one of the l_i to any value between zero and l, but keep the other l_j fixed, then we will still produce a valid configuration; hence the integration over the allowed range for each l_i. The endpoint of the clump is determined by requiring that there be no particle closer than l to it, hence the final exponential on the right-hand side of the square brackets.

In what follows, it will prove convenient to define

$$b \equiv 1 - \exp(-\bar{n}\, l). \quad (2)$$

Initially $l = 0$, so equation (1) says that all clumps have size $m = 1$ initially. At later times, $l \to \infty$, and the clump distribution tends to an exponential. In the process, b changes from zero to unity.

The fraction of mass which is in m clumps is

$$f(m, b) = \frac{m\eta(m, b)}{\Sigma m\eta(m, b)} = (1 - b)\, m\eta(m, b). \quad (3)$$

The number density of m-clumps is the average density $\bar{n}$ times $f(m, b)/m$, and is often called the universal mass function.

Consider an m-clump at t_2 when the link-length was $l_2 \geq l_1$. At $t_1 \leq t_2$ the link-length was shorter, and so not all m particles would have been "friends-of-friends." In other words, at $t_1 \leq t_2$, the m-clump was partitioned into smaller subclumps. A modification of the argument above allows us to write down the distribution of subclump sizes. Namely,

$$p(n_1, ..., n_m; l_1|m; l_2) = n! \frac{\eta(n, l_{21})}{\eta(m, l_2)} \prod_{i=1}^{m} \frac{\eta(i, l_1)^{n_i}}{n_i!}, \qquad (4)$$

where n_i denotes the number of subclumps which contain i particles; the total number of particles is $\Sigma\, in_i = m$ and they are partitioned into $\Sigma\, n_i = n$ clumps. The $n!$ factor comes from not caring about the order of the subclumps, the factor of $\eta(m,l_2)$ in the denominator comes from the fact that we know we have an m-clump at t_2, and $\eta(n, l_{21} \equiv l_2 - l_1)$ comes from noting that the distance between the right end of one subclump and the left end of the next subclump must be greater than l_1 but less than l_2. This gives a factor of $\exp(-\bar{n}\, l_1) - \exp(-\bar{n}\, l_2)$, which we write as $\exp(-\bar{n}\, l_1)[1 - e^{-\bar{n}(l_2 - l_1)}\}]$, for each subclump except the rightmost, for which the factor is simply $\exp(-\bar{n}\, l_2)$. The term in the product sign is just the probability of having the correct set of subclumps, and assuming that if subclumps that contain the same number of particles are exchanged, the partition of m is unchanged.

Because $\eta(i, l_1)$ depends on i only in the exponent, equation (4) can be simplified considerably:

$$p(n_1, ..., n_m | m) = \left(\frac{n!}{\prod_{i=1}^{m} n_i!}\right)\left(\frac{b_2 - b_1}{b_2}\right)^{n-1}\left(\frac{b_1}{b_2}\right)^{m-n}, \qquad (5)$$

where the b_i are defined by equation (2). This partition function contains all the information required for quantifying how different the various trees in the forest of possible merger histories of an m-clump are. For example, the probability that an m-clump was in n pieces at $t_1 < t_2$ is given by summing over the set $\pi(n|m)$ of partitions of m which have n parts:

$$p(n; b_1|m; b_2) = \sum_{\pi(n|m)} p(n_1, ..., n_m|m) = \binom{m-1}{n-1}\left(\frac{b_1}{b_2}\right)^{m-n}\left(\frac{b_2 - b_1}{b_2}\right)^{n-1}; \qquad (6)$$

the number of subclumps follows a Binomial distribution.

The average fraction of M which is in m-subclumps at b_1 is

$$f(m; b_1|M; b_2) = \sum_{\pi(M)} \frac{mn_m}{M} p(n_1, ..., n_M|M). \qquad (7)$$

The conditional mass function is related to this fraction similarly to how the unconditional mass function is related to $f(m, b)$: namely, $\mathcal{N}(m, b_1|M, b_2) = (M/m) f(m, b_1|M, b_2)$.

It is useful to rewrite the partition function (equation 4) above as follows. Let $p(S_n = m)$ denote the probability that the sum of n independent variables each distributed according to equation (1) equals m. Then

$$p(S_n = m; b) = \binom{m-1}{n-1}(1-b)^n b^{m-n}. \tag{8}$$

This, with equation (6) for the probability of having m subclumps, allows one to verify that equation (4) also equals

$$p(n_1, \ldots, n_m | m) = \frac{p(n; b_1|m; b_2)}{p(S_n = m; b_1)} n! \prod_{i=1}^{m} \frac{\eta(i, l_1)^{n_i}}{n_i!}. \tag{9}$$

Thus, the partition function expresses the fact that, other than the requirement that the sum of the masses of the subclumps should equal the mass of the parent halo, there are no additional correlations between subclumps.

All the above was worked out for the special case of a Poisson distribution of points on a line. To generalize these results to two (or d) dimensions we must be able to compute the area (volume) of interestection of n circles (d-dimensional spheres). Expressions for these quantites are in Kratky [1], [2]. Rather than showing the results of doing this here, I will now turn to another model of hierarchical clustering.

2.2. Gravity and a Critical Density for Collapse

The link-length model above was useful for illustrating how one might write expressions for the merger histories of clumps. It is a bad model for gravity for the following reason. The $m - 1$ separations between the m particles of an m-clump may all be a small fraction of the critical link-length l, but it is also possible that they are all a substantial fraction of l. As a result, the model allows m-clumps to have range of sizes. The model says that the gravitational influence of a clump extends over the region it occupies plus $2l$ (from the two endpoints). Because m-clumps come in a range of sizes, this would indicate that the gravitational influence of some m-clumps extends over a greater range than others. One might have thought, however, that it is the mass of a clump that determines the range of its influence and not the space it occupies.

To allow for this, we must modify the criterion used for identifying clumps. Rather than using only the list of initial separations, we use the initial densities. Assume that an initial region has collapsed to form a clump if the density within it exceeds a certain threshold value. In addition, assume that when a region collapses, all the particles within the region remain within it. This means that one is interested in finding regions that are isolated in the following sense: the points within the region must be sufficiently close to each other that the density within the region exceeds the critical value, but this set of points must be sufficiently isolated from any other set so that the density within any larger region containing the points is less than the critical value.

In this case, the probability a clump contains m particles is

$$\eta(m, b) = \frac{(mb)^{m-1} e^{-mb}}{m!} \quad \text{where} \quad b = \frac{1}{1 + \delta_c} \quad \text{and} \quad m \geq 1 \tag{10}$$

where $1 + \delta_c$ denotes the ratio of the critical density required for collapse to the average background density [3]. In this model, we assume that $\delta_c(t) \gg 1$ initially, and that it decreases with time. Notice that this distribution differs from equation (1) at low masses. At late times, this distribution tends to one that astronomers associate with Press and Schechter [4], rather than the simple exponential distribution of the previous subsection. The forest of merger trees associated with this model is given by inserting equation (10) in (9), using the distribution $p(S_n = m, b)$ associated with equation (10) rather than equation (8), and using the binomial distribution of subclumps (equation 6) [5].

In both this and the previous model, cosmology only enters when translating the critical link-length or density to cosmological time. The order in which mergers happen is the same for all cosmologies; it depends only on the initial distribution of density fluctuations. This is a powerful and simplifying idealization which is in good qualitative agreement with numerical simulations of clustering from cold, initially Gaussian density fluctuation fields.

The mass function and merger histories associated with equation (10) are in quantitative agreement with simulations, whereas the model associated with equation (1) is not. The quantitative agreement between the model description of the forest of merger histories and the simulations shows that any additional correlations between subclumps, other than those required by mass conservation, must be small.

2.3. Random Walks and the Excursion Set Approach

The mass function (equation 10) above can be derived by rephrasing the critical density requirement slightly. If one imagines computing the density ρ in concentric spheres centered on a randomly chosen particle in a Poisson distribution, then $\rho(v)$ will execute a random walk as v increases. The requirement that the density exceed a certain value at v, but be less than this value for all $V > v$, means that the problem of computing the mass function can be cast in terms of a barrier crossing problem associated with random walks [6]. The continuum limit of this process has been studied by Bond et al. [7], and has come to be called the excursion set model of the clump mass function. This approach allows one to compute the conditional mass function as well—it is the continuum limit of the distribution one gets by inserting equation (10) in equation (7)—but the full partition function of merger histories associated with this model has not yet been worked out (Sheth and Pitman [8] discuss a special case in which the merger history forest can be solved for). Progress can be made, however, if one uses the idea that, other than mass conservation, there are no additional correlations to account for. The resulting model is in good agreement with simulations [9].

I have not seen a random walk derivation of the exponential distribution associated with the link-length model I presented earlier.

2.4. Binary Merger Models

The models previously described are also solutions of the Smoluchowski binary merger model:

$$\frac{dn(m,t)}{dt} = \sum_{i=1}^{m-1} \frac{K(i, m-i)}{2} n(i,t) n(m-i,t) - n(m,t) \sum_{i>0} K(m,i) n(i,t). \tag{11}$$

The expression above expresses the fact that the number of clumps of mass m increases if smaller clumps merge with each other to form an m-clump, and it decreases because m-clumps are themselves merging with other clumps. Note that there is no fragmentation in this model—the destruction rate which is the second term on the right-hand side is a consequence of mergers, not fragmentation.

If we set $n(m, t) = (1 - b)\eta(m, b)$, where $b \equiv 1 - \exp(-t)$, then equation (1) solves the case in which $K(i, j)$ is a constant, independent of both i and j. It has been used to describe the growth of linear polymers. Our second model, equation (10), is the solution to $K = i + j$; it is associated with the growth of branched polymers (e.g., Sheth and Pitman [8]).

One of the virtues of writing these models using Smoluchowski's equation is that it shows clearly how the formation rate of m-clumps evolves with time. It is given by the first term on the right-hand side of equation (11). It is a simple matter to verify that the continuum limit of this expression equals that which was recently found by Percival and Miller [10] from the random walk approach.

2.5. Peaks in Gaussian Random Fields

Another model for the clump distribution is to suppose that clumps are associated with peaks in the initial density field. Following Bardeen *et al.* [11], one often smooths the initial density fluctuation field with a filter of scale R, and then identifies peaks in the smoothed field. In this case, the density of peaks of height ν is

$$n(\nu) d\nu = \frac{\exp(-\nu/2)}{(2\pi)^2 R_*^3} G(\gamma, \gamma \nu^{1/2}) d\nu \tag{12}$$

where $\nu = \delta_c^2/\sigma_0^2(R)$, $R_* = \sqrt{3}\sigma_1/\sigma_0$, $\gamma(R) = \sigma_1^2/\sigma_0\sigma_2$, and σ_0, σ_1 and σ_2 depend on the shape of the power spectrum of the initial density fluctuation field [11, Sect. IV] and

$$G(\gamma, y) = \int_0^\infty dx\, f(x) \frac{\exp[-(x-y)^2/2(1-\gamma^2)]}{\sqrt{2\pi(1-\gamma^2)}}$$

with $f(x)$ given by equation (A19) in Bardeen et al. [11]. If we define

$$f(\nu)\, d\nu \equiv (m/\bar{\rho})\, n(\nu)\, d\nu, \tag{13}$$

and use the fact that the mass under a Gaussian filter is $m = \bar{\rho}\,(2\pi)^{3/2} R^3$, then we have a quantity that one might interpret as the fraction of mass which is in peaks of height ν.

Unfortunately, this is not the sort of quantity we can compare with a mass function of collapsed clumps. In simulations, clumps have a range of masses, whereas in this picture all peaks have the same mass m whatever their height ν. Although it is tempting to identify the higher peaks with the more massive objects, the expression above does not show how to do this self-consistently.

If, instead, we smoooth the density field with a range of filter sizes R, and identify collapsed objects with peaks of height $\delta_c/\sigma_0(R)$, then because $\sigma_0(R)$ decreases as $R \propto m^{1/3}$ increases, we have a model in which massive objects are associated with higher peaks. In this case, the associated mass function of collapsed peaks is given by

$$\nu f(\nu) = \frac{\exp(-\nu/2)}{\sqrt{2\pi}} \left(\frac{R}{R_*}\right)^3 \frac{H(\gamma, \gamma\nu^{1/2})}{3} \frac{R^2 \sigma_2(R)}{\sigma_0(R)} \frac{dm/m}{d\nu/\nu} \tag{14}$$

where

$$H(\gamma, y) = \int_0^\infty dx\, x f(x) \frac{\exp[-(x-y)^2/2(1-\gamma^2)]}{\sqrt{2\pi(1-\gamma^2)}},$$

and we have again set $m/\bar{\rho} = (2\pi)^{3/2} R^3$. At large ν, $H \approx \gamma \nu^{1/2}\, G$, and this expression is the same as the previous one. At smaller ν, however, this expression differs from equation (12).

If the initial spectrum of density fluctuations was a power law, $P(k) \propto k^n$, then equation (14) for the mass function associated with peaks becomes

$$\nu f(\nu) = \sqrt{\frac{\nu}{2\pi}} \exp\left(-\frac{\nu}{2}\right) \frac{H(\gamma, \gamma\nu^{1/2})}{\nu^{1/2}} \frac{(5+n)^2}{12\sqrt{6(3+n)}}. \tag{15}$$

Note that this expression explicitly depends on the shape of the power spectrum.

This should be compared with the continuum limit of equation (10),

$$\nu f(\nu) = \sqrt{\frac{\nu}{2\pi}} \exp\left(-\frac{\nu}{2}\right), \tag{16}$$

which is often called the Press–Schechter formula. Note that, when expressed as a function of ν rather than m, this mass function is the same for all $P(k)$.

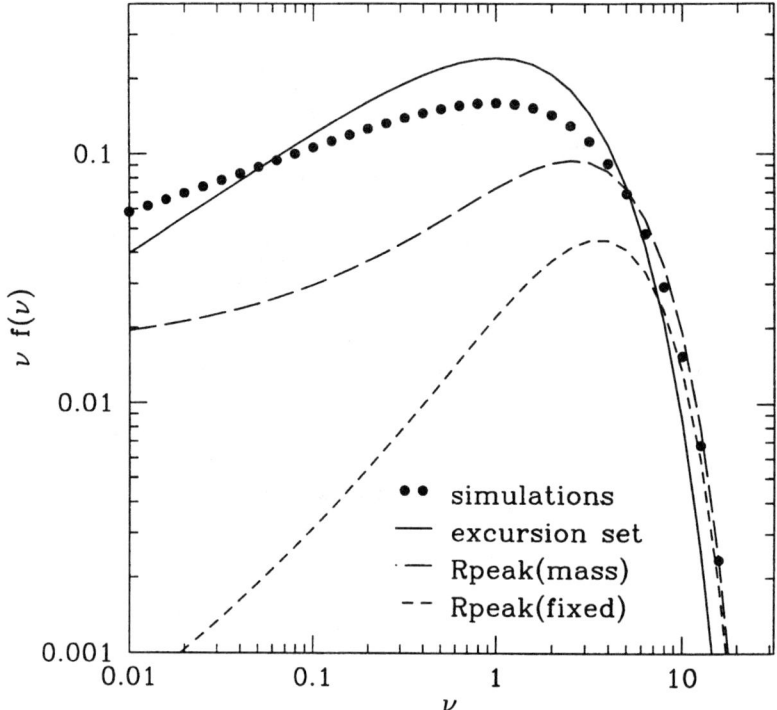

FIGURE 1. Various formulae for the mass function of collapsed objects. Filled circles show the mass function measured in numerical simulations of clustering in a cold dark matter dominated universe. Solid curve shows the excursion set spherical collapse formula, short dashed line shows the result of using a fixed smoothing scale to define peaks, and then simply counting peaks as a function of height, and long dashed line shows the result of assuming that more massive objects are associated with peaks on large smoothing scales, whereas less massive objects are peaks on smaller scales.

The mass function which actually fits numerical simulations is better approximated by

$$\nu f(\nu) = A(p)(1 + (a\nu)^{-p})\sqrt{\frac{a\nu}{2\pi}}\exp\left(-\frac{a\nu}{2}\right), \qquad (17)$$

where $a \approx 0.7$, $p = 0.3$ and $A(p)$ is determined by requiring that $\int d\nu\, f(\nu) = 1$; $A \approx 0.322$ [12]. Like the Press–Schechter function, this mass function also has no dependence on the shape of $P(k)$.

These various mass functions are shown in FIGURE 1. Whereas the Press–Schechter, excursion set, critical collapse density, binary merger model, is not in quantitative agreement with the mass function of collapsed objects in the simulations, it is at least qualitatively consistent. The peaks model fits the simulations rather well at high masses (large ν) but is not so accurate at smaller masses. (I set $n = -1.5$ in the peaks formula which is about the right value for ΛCDM models.)

Equation (16) is associated with the assumption that clumps form from a spherical collapse (e.g., [4] and [7]). Modifying the excursion set argument to allow for ellipsoidal collapse is relatively straightforward [13]. In essence, the spherical model has $\delta_{sc} \approx 1.686$ with no m dependence for the critical collapse density, whereas ellipsoidal collapse has $\delta_{ec}(m)$. At large m, $\delta_{ec}(m) \to \delta_{sc}$; this simply reflects the fact that only the most massive clusters are spherical. Accounting for the difference between ellipsoidal and spherical collapse appears to increase agreement between the excursion set predictions and the simulations.

Modifying the peaks mass function to include the effects of ellipsoidal collapse can be done by using the mass dependent $\delta_{ec}(m)$ when identifying peaks; this is relatively straightforward and the results are not shown here. Extending the peaks model to provide a description of the forest of merger histories remains an open problem (but see Manrique et al. [14] for an initial attempt to do this). Also, I have not seen any discussion of how one might include a mass dependent $\delta_{ec}(m)$ into the Smoluchowski binary merger model of the mass function.

3. THE SPATIAL DISTRIBUTION OF MASS

In this section, two approaches to modeling how the distribution of the density field evolves in the nonlinear regime are described. Both approaches are quite different from the linear and higher-order perturbation theory approaches (e.g., Scoccimarro [15, this volume]) that solve the equations of motion to make their estimates.

The essential idea is that if all the mass is in collapsed objects, then one can describe the distribution of the mass by making models of the number and spatial distribution of clumps, and of the internal distribution of mass within clumps [16], [17]. Numerical simulations show that the density run within a clump depends on clump mass; Navarro, Frenk, and White [18] have provided a simple fitting formula for this mass dependence. The previous section described models of the mass function of collapsed objects. All that remains is to build a model for the spatial distribution of clumps.

Mo and White [19] described how knowledge of the merger history tree of clumps allows one to describe the spatial distribution of clumps. They argued that the two-point correlation function of clumps can be computed if the second moment of the distribution of clump merger histories is known. At large separations, they argued that knowledge of only the conditional and unconditional clump mass functions was necessary for modeling the clump correlation functions. Sheth and Tormen [12] showed that, in fact, in this limit, knowledge of only the unconditional mass function is sufficient.

Mo, Jing, and White [20] used equation (16) to write estimates of the variance and higher-order moments of the large scale clump distribution. They also did this for the "fixed smoothing scale" peaks mass function of equation (12). The corresponding expressions for the peak mass function in equation (14) are obtained by replacing their equation (25) with:

$$h_k = \frac{(-1)^k}{k!} \frac{(\gamma v^{1/2})^k}{H(\gamma, \gamma v^{1/2})} \frac{\partial^k H(\gamma, \gamma v^{1/2})}{\partial y^k} \bigg|_{y = \gamma v^{1/2}}.$$

In the approach above, one solves for the distribution of the mass by writing estimates of the two-point and higher-order correlation functions. If one is interested in writing the probability that a randomly placed cell contains mass M, then one must compute the appropriate sum over all these correlation functions.

It is possible to build a model for this probability distribution function directly [21]. The idea is to modify slightly the random walk, excursion set model of the clump mass function. In this modified excursion set approach, the same model which yields equations (10) and (16) for the mass function says that the probability that a random cell contains mass m is *generalized Poisson* and *inverse Gaussian*, respectively. Remarkably, this inverse Gaussian distribution is also predicted by the perturbation theory-based model of Scoccimarro and Frieman [22].

The fundamental quantity in the excursion set model for the clump mass function is the distribution of first crossings, by Brownian motion random walks, of a barrier of fixed height $B = \delta_{sc}$. If one thinks of clumps as being objects which have collapsed to a very small size, then the clump mass function is like the distribution of mass in cells of vanishingly small size which are not empty. This allows one to generalize the model to the case in which the cells have some nonzero size. To do so, one must study the first crossing distribution of a barrier whose shape $B(M/V)$ depends on cell size V. If $f(M)dM$ denotes the probability that the first crossing of $B(M/V)$ happens at M, then

$$f(M)dM = \frac{M}{\rho V} p(M|V) \, dM, \tag{18}$$

where $f(M)dM$ denotes the probability that a cell of size V contains M [21]. Of course, the first crossing distribution, and so the associated probability that a cell contains mass M, both depend on the functional form of B. In 1998, I used the spherical collapse model to specify the shape of this function.

In the limit of large V, the associated distribution of first crossings of this barrier is well approximated by

$$f(M)dM \approx p(B) \, dB, \tag{19}$$

where B depends on M and V. If the initial distribution of fluctuations was Gaussian then we should set $p(x)$ to be Gaussian. If we then insert this in equation (18), we have a model for the large scale probability distribution of the mass. The result of doing this is very similar to the model of the nonlinear probability distribution derived by Fosalba and Gaztañaga [23].

Fosalba and Gaztañaga argued that their analysis could not be applied on small scales. Our excursion set model, however, can be used even on smaller scales. The only caveat is that the first crossing distribution is more complicated than the simple transformation of the Gaussian given above. In this sense, the excursion set model can be thought of as providing a simple way to extend the results presented in Fosalba and Gaztañaga to smaller, more nonlinear scales.

4. DISCUSSION

It has become common practice to announce that cosmology, as a subject, has matured. While this is welcome news, I fear that this maturity also signifies a sort of loss of innocence. Interest has shifted from simple models which capture the essence of the nonlinear physics of gravitational instability, to detailed numerical simulations of the growth of clustering, in which the problem is solved by brute force, sometimes with no net increase in physical insight—the exponential growth in computing power in recent years has not been accompanied by a corresponding increase in our understanding of how clustering evolves.

Some aspects of the first two simple models of the growth of hierarchical clustering described above are quite general; they are not restricted to the nonlinearities associated with gravitational instability in an expanding Universe.

ACKNOWLEDGMENTS

This work was supported by the DOE and NASA grant NAG 5-7092 at Fermilab.

REFERENCES

1. KRATKY, K.W. 1978. The area of intersection of n equal circular disks. J. Phys. A: Math. Gen. **11(6)**: 1017–1024.
2. KRATKY, K.W. 1981. Intersecting discs (and spheres) and statistical mechanics. I. Mathematical basis. J. Stat. Phys.**25(4)**: 619–634.
3. SHETH, R.K. 1995. Merging and hierarchical clustering from an initially Poisson distribution. Mon. Not. R. Astr. Soc. **276**: 796–824.
4. PRESS, W.H. & P. SCHECHTER. 1974. Formation of galaxies and clusters of galaxies by self-similar gravitational condensation. Astrophys. J. **187**: 425–438.
5. SHETH, R.K. 1996. Galton–Watson branching processes and the growth of gravitational clustering. Mon. Not. R. Astr. Soc. **281**: 1277–1289.
6. EPSTEIN, R. 1983. Proto-galactic perturbations. Mon. Not. R. Astr. Soc. **205**: 207–229
7. BOND, J.R., S. COLE, G. EFSTATHIOU & N. KAISER. 1991. Excursion set mass functions for hierarchical Gaussian fluctuations. Astrophys. J. **379**: 440–460.
8. SHETH, R.K. & J. PITMAN. 1997. Coagulation and branching process models of gravitational clustering. Mon. Not. R. Astr. Soc. **289**: 66–82.
9. SHETH, R.K. & G. LEMSON. 1999. The forest of merger history trees associated with the formation of dark matter haloes. Mon. Not. R. Astr. Soc. **305**: 946–956.
10. PERCIVAL, W. & L. MILLER. 1999. Cosmological evolution and hierarchical galaxy formation. Mon. Not. R. Astr. Soc. **309**: 823–832.
11. BARDEEN, J.M., J.R. BOND, N. KAISER & A.S. SZALAY. 1986. The statistics of peaks of Gaussian random fields. Astrophys. J. **304**: 15–61.
12. SHETH, R.K. & G. TORMEN. 1999. Large-scale bias and the peak background split. Mon. Not. R. Astr. Soc. **308**: 119–126.
13. SHETH, R.K., H. MO & G. TORMEN. 2001. Ellipsoidal collapse and an improved model for the number and spatial distribution of dark matter haloes. Mon. Not. R. Astr. Soc. In press.

14. MANRIQUE, A., A. RAIG, J.-M. SOLANES, G. GONZÀLEZ-CASADO, P. STEIN & E. SALVADO-SOLÉ. 1998. The effects of the peak-peak correlation on the peak model of hierarchical clustering. Astrophys. J. **499:** 548–554.
15. SCOCCIMARRO, R. 2001. A new angle on gravitational clustering. Ann. N.Y. Acad. Sci. **927:** this volume.
16. NEYMAN, J. & E.L. SCOTT. 1952. A theory of the spatial distribution of galaxies. Astrophys. J. 116: 144–163.
17. SCHERRER, R.J. & E. BERTSCHINGER. 1991. Statistics of primordial density perturbations from discrete seed masses. Astrophys. J. **381:** 349–360.
18. NAVARRO, J.F., C.S. FRENK & S.D.M. WHITE. 1997. A universal density profile from hierarchical clustering. Astrophys. J. **490:** 493–508.
19. MO, H. & S.D.M. WHITE. 1996. An analytic model for the spatial clustering of dark matter haloes. Mon. Not. R. Astr. Soc. **282:** 347–361.
20. MO, H., Y. JING & S.D.M. WHITE. 1997. High-order correlations of peaks and haloes: a step towards understanding galaxy biasing. Mon. Not. R. Astr. Soc. **284:** 189–201.
21. SHETH, R.K. 1998. An excursion set model for the distribution of dark matter and dark matter haloes. Mon. Not. R. Astr. Soc. **300:** 1057–1070.
22. SCOCCIMARRO, R. & J. FRIEMAN. 1999. Hyperextended cosmological perturbation theory: Predicting nonlinear clustering amplitudes}, Astrophys. J. **520:** 35–44
23. FOSALBA, P. & E. GAZTAÑAGA. 1998. Cosmological Perturbation Theory and the Spherical Collapse Model. I. Gaussian Initial Conditions. Mon. Not. R. Astr. Soc. **301:** 503–523.

A New Angle on Gravitational Clustering

ROMÁN SCOCCIMARRO

Institute for Advanced Study, School of Natural Sciences, Einstein Drive, Princeton, New Jersey 08540

ABSTRACT: A new approach to gravitational instability in large-scale structure is described, where the equations of motion are written and solved as in field theory in terms of Feynman diagrams. The basic objects of interest are the propagator (which propagates solutions forward in time), the vertex (which describes nonlinear interactions between waves) and a source with prescribed statistics which describes the effect of initial conditions. We show that loop corrections renormalize these quantities, and discuss applications of this formalism to a better understanding of gravitational instability and to improving nonlinear perturbation theory in the transition to the nonlinear regime. We also consider the role of vorticity creation due to shell-crossing and show using N-body simulations for which at small (virialized) scales the velocity field reaches equipartition, that is, the vorticity power spectrum is about twice the divergence power spectrum.

KEYWORDS: Gravitational instability; Large-scale structure of the Universe

1. STANDARD FORMULATION OF GRAVITATIONAL INSTABILITY

Assuming the initial velocity field is irrotational, gravitational instability can be described completely in terms of the density field and the velocity divergence, $\theta \equiv \nabla \cdot \mathbf{v}$. Defining the conformal time $\tau = \int dt/a$ and the conformal expansion rate $\mathcal{H} \equiv d\ln a/d\tau$, the equations of motion in Fourier space become

$$\frac{\partial \tilde{\delta}(\mathbf{k})}{\partial \tau} + \tilde{\theta}(\mathbf{k}) = -\int d^3k_1 d^3k_2 [\delta_D] \alpha(\mathbf{k},\mathbf{k}_1) \tilde{\theta}(\mathbf{k}_1) \tilde{\delta}(\mathbf{k}_2), \qquad (1)$$

$$\frac{\partial \tilde{\theta}(\mathbf{k})}{\partial \tau} + \mathcal{H}\tilde{\theta}(\mathbf{k}) + \frac{3}{2}\Omega \mathcal{H}^2 \tilde{\delta}(\mathbf{k}) = -\int d^3k_1 d^3k_2 [\delta_D] \beta(\mathbf{k}_1,\mathbf{k}_2) \tilde{\theta}(\mathbf{k}_1) \tilde{\theta}(\mathbf{k}_2), \qquad (2)$$

where $[\delta_D] = \delta_D(\mathbf{k} - \mathbf{k}_{12})$, k is a comoving wavenumber, and

$$\alpha(\mathbf{k},\mathbf{k}_1) \equiv \frac{\mathbf{k} \cdot \mathbf{k}_1}{k_1^2}, \quad \beta(\mathbf{k}_1,\mathbf{k}_2) \equiv \frac{k^2(\mathbf{k}_1 \cdot \mathbf{k}_2)}{2k_1^2 k_2^2}. \qquad (3)$$

Address for correspondence: Roman Scoccimarro, Department of Physics, New York University, 4 Washington Place, New York, NY 10003. Voice: 212/998-7649; fax: 212/995-4016.
scoccima@physics.nyu.edu

Equations (1) and (2) are valid in an arbitrary homogeneous and isotropic universe, which evolves according to the Friedmann equations:

$$\frac{\partial \mathcal{H}(\tau)}{\partial \tau} = -\frac{\Omega}{2}\mathcal{H}^2(\tau) + \frac{\Lambda}{3}a^2(\tau), \tag{4}$$

$$(\Omega - 1)\mathcal{H}^2(\tau) = k - \frac{\Lambda}{3}a^2(\tau), \tag{5}$$

where Λ is the cosmological constant, the spatial curvature constant $k = -1, 0, 1$ for $\Omega_{\text{tot}} < 1$, $\Omega_{\text{tot}} = 1$, and $\Omega_{\text{tot}} > 1$, respectively, and $\Omega_{\text{tot}} \equiv \Omega + \Omega_\Lambda$, with $\Omega_\Lambda \equiv \Lambda a^2/(3\mathcal{H}^2)$. For $\Omega = 1$, the perturbative growing mode solutions are given by

$$\tilde{\delta}(k) = \sum_{n=1}^{\infty} a^n(\tau)\delta_n(k), \tag{6}$$

$$\tilde{\theta}(k) = \mathcal{H}(\tau)\sum_{n=1}^{\infty} a^n(\tau)\theta_n(k), \tag{7}$$

Modelling the matter as pressureless nonrelativistic "dust," an appropriate description for cold dark matter before shell crossing, the fluid equations of motion determine $\delta_n(k)$ and $\theta_n(k)$ in terms of the linear fluctuations,

$$\delta_n(k) = \int d^3q_1 \ldots \int d^3q_n \, [\delta_D]_n \, F_n^{(s)}(q_1, \ldots, q_n) \, \delta_1(q_1) \ldots \delta_1(q_n) \tag{8}$$

$$\theta_n(k) = -\int d^3q_1 \ldots \int d^3q_n \, [\delta_D]_n \, G_n^{(s)}(q_1, \ldots, q_n) \, \delta_1(q_1) \ldots \delta_1(q_n) \tag{9}$$

where $[\delta_D]_n \equiv \delta_D(k - q_1 - \ldots - q_n)$. The functions $F_n^{(s)}$ and $G_n^{(s)}$ are constructed from the mode coupling functions $\alpha(k, k_1)$ and $\beta(k_1, k_2)$ by a recursive procedure [1],

$$F_n(q_1, \ldots, q_n) =$$

$$\sum_{m=1}^{n-1} \frac{G_m(q_1, \ldots, q_m)}{(2n+3)(n-1)}[(2n+1)\alpha(k, k_1)G_{n-m}(q_{m+1}, \ldots, q_n)]$$

$$+ 2\beta(k_1, k_2)G_{n-m}(q_{m+1}, \ldots, q_n) \tag{10}$$

(where $k_1 \equiv q_1 + \ldots + q_m$, $k_2 \equiv q_{m+1} + \ldots + q_n$, $k \equiv k_1 + k_2$, and $F_1 = G_1 \equiv 1$). In Eqs.(8) and (9), $F_n^{(s)}$ and $G_n^{(s)}$ are the symmetrized version of F_n and G_n, respectively.

$$G_n(\boldsymbol{q}_1, \ldots, \boldsymbol{q}_n) = \sum_{m=1}^{n-1} \frac{G_m(\boldsymbol{q}_1, \ldots, \boldsymbol{q}_m)}{(2n+3)(n-1)} [3\alpha(\boldsymbol{k}, \boldsymbol{k}_1) F_{n-m}(\boldsymbol{q}_{m+1}, \ldots, \boldsymbol{q}_n)]$$
$$+ 2n\beta(\boldsymbol{k}_1, \boldsymbol{k}_2) G_{n-m}(\boldsymbol{q}_{m+1}, \ldots, \boldsymbol{q}_n) \quad (11)$$

From these perturbative solutions a number of important results have been derived in the literature, most of them regarding the tree-level (leading-order) behavior of correlation functions, e.g., references [1]–[4]. Loop calculations have been attempted only in some particular cases [5]–[7]; although in the spherical collapse approximation a number of useful results have been obtained [8], [9].

2. FIELD THEORY APPROACH

2.1. Integral Form of the Equations of Motion

The equations of motion can be rewritten in a more symmetric form by defining a two-component "vector" $\Psi_{a(\boldsymbol{k},z)}$, where $a = 1,2$, $z \equiv \ln a$, and:

$$\Psi_a(\boldsymbol{k},z) \equiv ((\delta(\boldsymbol{k}, z), -\theta(\boldsymbol{k}, z)/\mathcal{H}), \quad (12)$$

which for $\Omega = 1$ leads to the following equations of motion (we henceforth use the convention that repeated Fourier arguments are integrated over)

$$\partial_z \Psi_a(\boldsymbol{k},z) + \Omega_{ab} \Psi_b(\boldsymbol{k}, z) = \gamma_{abc}(\boldsymbol{k}, \boldsymbol{k}_1, \boldsymbol{k}_2) \Psi_b(\boldsymbol{k}_1, z) \Psi_c(\boldsymbol{k}_2, z) \quad (13)$$

where

$$\Omega_{ab} \equiv \begin{bmatrix} 0 & -1 \\ -3/2 & 1/2 \end{bmatrix} \quad (14)$$

and γ_{abc} is a matrix given by

$$\gamma_{121}(\boldsymbol{k}, \boldsymbol{k}_1, \boldsymbol{k}_2) = \delta_D(\boldsymbol{k} - \boldsymbol{k}_{12}) \, \alpha(\boldsymbol{k}, \boldsymbol{k}_1) \quad (15)$$

$$\gamma_{222}(\boldsymbol{k}, \boldsymbol{k}_1, \boldsymbol{k}_2) = \delta_D(\boldsymbol{k} - \boldsymbol{k}_{12}) \, \beta(\boldsymbol{k}_1, \boldsymbol{k}_2) \quad (16)$$

and zero otherwise. The somewhat complicated expressions for the PT kernels recursion relations in the previous section can be easily derived in this formalism. The perturbative solutions read

$$\Psi_a(\boldsymbol{k}, z) = \sum_{n=1}^{\infty} e^{nz} \psi_a^{(n)}(\boldsymbol{k}), \quad (17)$$

which leads to

Now, let $\sigma_{ab}^{-1}(n) \equiv n\delta_{ab} + \Omega_{ab}$, then we have:

$$(n\delta_{ab} + \Omega_{ab})\psi_b^{(n)}(k) = \gamma_{abc}(k, k_1, k_2) \sum_{m=1}^{n-1} \psi_b^{(n-m)}(k_1)\psi_c^{(m)}(k_2). \quad (18)$$

$$\psi_a^{(n)}(k) = \sigma_{ab}(n) \gamma_{bcd}(k, k_1, k_2) \sum_{m=1}^{n-1} \psi_c^{(n-m)}(k_1)\psi_d^{(m)}(k_2), \quad (19)$$

where

$$\sigma_{ab}(n) = \frac{1}{(2n+3)(n-1)}\begin{bmatrix} 2n+1 & 2 \\ 3 & 2n \end{bmatrix}. \quad (20)$$

Equation (19) is the equivalent to the Goroff *et al.* [1] recursion relations, Eqs. (10) and (11), for the nth-order Fourier amplitude solutions $\psi_a^{(n)}(k)$. It turns out to be convenient to write down the equation of motion Eq. (13) in integral form. Laplace transformation in the variable z leads to:

$$\sigma_{ab}^{-1}(\omega) \Psi_b(k, \omega) = \phi_a(k) + \gamma_{abc}(k, k_1, k_2) \oint \frac{d\omega_1}{2\pi i}\Psi_b(k_1, \omega_1)\Psi_c(k_2, \omega - \omega_1), \quad (21)$$

where $\phi_a(k)$ denote the initial conditions, that is $\Psi_a(k, z=0) \equiv \phi_a(k)$. Multiplying by the matrix σ_{ab}, and performing the inversion of the Laplace transform we finally get

$$\Psi_a(k, z) = g_{ab}(z) \phi_b(k) + \int_0^z dz' g_{ab}(z-z') \gamma_{bcd}(k, k_1, k_2) \Psi_c(k_1, z')\Psi_d(k_2, z') \quad (22)$$

where the *linear propagator* $g_{ab}(z)$ is defined as ($c > 1$ to pick out the standard retarded propagator [10])

$$g_{ab}(z) = \oint_{c-i\infty}^{c+i\infty} \frac{d\omega}{2\pi i}\sigma_{ab}(\omega) e^{\omega z} = \frac{e^z}{5}\begin{bmatrix} 3 & 2 \\ 3 & 2 \end{bmatrix} - \frac{e^{-3z/2}}{5}\begin{bmatrix} -2 & 2 \\ 3 & -3 \end{bmatrix}, \quad (23)$$

for $z \geq 0$, whereas $g_{ab}(z) = 0$ for $z < 0$ due to causality, $g_{ab}(z) \to \delta_{ab}$ as $z \to 0^+$. The first term in Eq. (23) represents the propagation of linear growing mode solutions, where the second corresponds to the decaying modes propagation. If we assume that the initial conditions are set in the growing mode, then $\phi_a(k) = \delta_1(k)(1,1)$ and the second term in Eq. (23) does not contribute upon contraction with $\phi_b(k)$. Consistent with this, we can neglect subdominant time dependences in the nonlinear term in Eq. (22), which amounts to setting the initial conditions at $z = -\infty$. Then, the equations of motion in integral form reduce to:

$$\Psi_a(k, z) = e^z \phi_a(k) + \int_{-\infty}^z dz' g_{ab}(z-z') \gamma_{bcd}(k, k_1, k_2) \Psi_c(k_1, z')\Psi_d(k_2, z'), \quad (24)$$

As it stands, this integral equation can be thought of as an equation for $\Psi_a(k, z)$ in the presence of an "external source" $\phi_a(k)$ with prescribed statistics. In particular, if we assume that the initial conditions are Gaussian; then the statistical properties of $\phi_a(k)$ are completely characterized by its two-point correlator

$$\langle \phi_a(k) \phi_b(k') \rangle = \delta_D(k + k') u_{ab} P(k), \tag{25}$$

where $P(k)$ denotes the initial power spectrum of density fluctuations and $u_{ab} = 1$ for growing-mode initial conditions. From Eq. (24), it is easy to verify that the ansatz in Eq. (17) leads to the recursion relations in Eq. (19).

Equation (22) has a simple interpretation. The first term corresponds to the linear propagation from the initial conditions, whereas the second term contains information on nonlinear interactions (mode–mode coupling). This corresponds to all the interactions between waves that happen at time z' (with $0 \leq z' \leq z$) characterized by γ_{bcd} and then propagated forward in time from z' to z by the propagator $g_{ab}(z - z')$. We can also write down the general solution as a perturbation series,

$$\Psi_a^{(n)}(k, z) = \int \delta_D(k - k_{1...n}) \mathcal{F}_a^{(n)}(k_1, ..., k_n; z) \delta_1(k_1) ... \delta_1(k_n), \tag{26}$$

where the kernels satisfy the usual recursion relations

$$\mathcal{F}_a^{(n)}(z) = \sum_{m=1}^{n} \int_0^z ds\, g_{ab}(z - s) \gamma_{bcd}(k, k_1, k_2) \mathcal{F}_c^{(m)}(s) \mathcal{F}_d^{(n-m)}(s) \tag{27}$$

Interactions modify the linear propagator, leading to *propagator renormalization*, so that the nonlinear propagator defined by

$$G_{ab}(k, z) \delta_D(k - k') \equiv \left\langle \frac{\delta \Psi_a(k, z)}{\delta \phi_b(k')} \right\rangle_c, \tag{28}$$

reads

$$G_{ab}(k, z) = g_{ab}(z) + \sum_{n=1}^{\infty} A_n(z) \int d^3 q_1 P_1 ... d^3 q_n P_n \frac{\partial \overline{\mathcal{F}}_a^{2(n+1)}}{\partial u_b}, \tag{29}$$

where $\overline{\mathcal{F}}_a^{2(n+1)} = \mathcal{F}_a^{(2n+1)}(k, q_1, -q_1, ..., q_n, -q_n)$, $A_n(z) \equiv (2n-1)!! \exp[(2n+1)z]$, and we defined $\phi_b \equiv (u_1, u_2) \delta_1(k)$. Similarly the vertex is renormalized by nonlinear interactions as well,

$$\Gamma_{abc}(k_1, k_2, z) \delta_D(k - k_{12}) \equiv G_{bd}^{-1} G_{ce}^{-1} \left\langle \frac{\delta^2 \Psi_a(k, z)}{\delta \phi_d(k_1) \delta \phi_e(k_2)} \right\rangle_c, \tag{30}$$

and thus $\Gamma_{abc} = \gamma_{abc}$ + corrections.

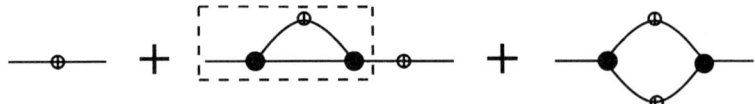

FIGURE 1. Power spectrum diagrams up to one-loop. The first term denotes the linear contribution, the two remaining terms denote the one-loop correction. The factor enclosed by dashed lines denotes propagator renormalization.

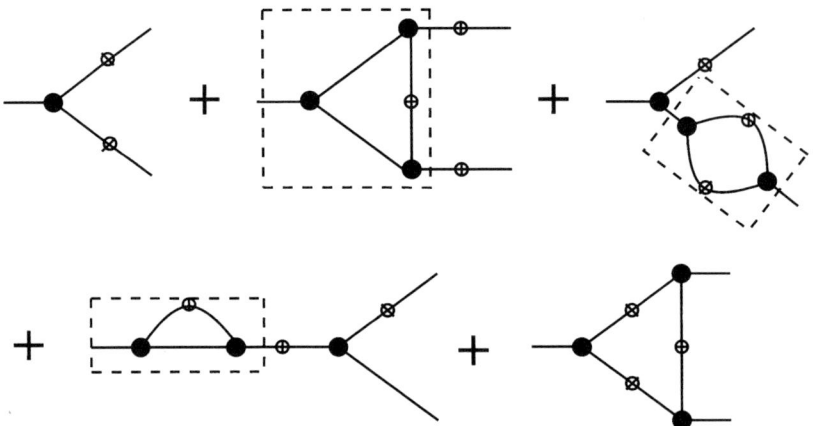

FIGURE 2. Bispectrum diagrams up to one-loop. The first terms denotes the tree contribution, the four remaining terms the one-loop correction. The first factor enclosed by dashed lines denotes vertex renormalization, the second corresponds to the irreducible one-loop power spectrum, the third denotes propagator renormalization. The last term gives the irreducible one-loop bispectrum.

The calculation of correlation functions can be written down in terms of Feynman diagrams (FIGS. 1 and 2). We assign a solid line to each propagator, Eq. (23), a crossed circle represents the two-point correlator in the initial conditions, Eq. (25), and a solid circle represent the vertex, Eqs. (15) and (16). In this representation, loop corrections can be divided into two general classes, those which renormalize the propagator, and those which renormalize the vertex. For example, in the calculation of the one-loop power spectrum there are two contributions, $P^{(1)} = \langle \delta_1 \delta_3 + \delta_2 \delta_2 \rangle_c$ (where δ_i denotes the ith-order solution in PT); the $\langle \delta_1 \delta_3 \rangle_c$ term corresponds to renormalizing the propagator (first one-loop term in FIG. 1), whereas the $\langle \delta_2 \delta_2 \rangle_c$ term denotes the irreducible one-loop power spectrum (second one-loop term in FIG. 1). By irreducible, we mean that this contribution cannot be separated into two connected diagrams by cutting one internal propagator line, unlike the $\langle \delta_1 \delta_3 \rangle_c$ contribution.

For the bispectrum, the 4 one-loop terms can be divided in a similar fashion. The $\langle \delta_4 \delta_1^2 \rangle_c$ corresponds to vertex renormalization (first one-loop term in FIG. 2), the $\langle \delta_3 \delta_2 \delta_1 \rangle_c$ correspond to power spectrum (second one-loop term in FIG. 2) and propagator renormalization (third one-loop term in FIG. 2), and the $\langle \delta_2^3 \rangle_c$ gives the irreducible one-loop bispectrum (last term in FIG. 2). This formalism can also be extended to include non-Gaussian initial conditions, see Scoccimarro [10] for a general discussion and the specific example of Zel'dovich approximation initial conditions, relevant to transients in N-body simulations.

3. ONE-LOOP PROPAGATOR AND THE NONLINEAR POWER SPECTRUM

As an example, we calculate the one-loop propagator, $G_{ab}^{(1)}$

$$G_{ab}^{(1)}(k, z) = g_{ab}(z) + \exp(3z)\int d^3q P(q) \frac{\partial \mathcal{F}_a^{(3)}}{\partial u_b}(k, q, -q; z) \quad (31)$$

If we take the $k \to 0$ limit, we find that (keeping only the fastest growing term)

$$G_{ab}^{(1)}(k, z) = g_{ab}(z) - k^2 \sigma_v^2 \exp(3z) \begin{bmatrix} 9/50 & 61/1050 \\ 3/25 & 61/1575 \end{bmatrix}, \quad (32)$$

where $\sigma_v^2 \equiv \int P(q) \, d^3q/q^2$. Since the correction is negative, this tends to make the nonlinear growth smaller than in linear theory, particularly for linearly decaying modes, which decay faster than in the linear case. The correction to g_{ab} can be rewritten in terms of a correction to Ω_{ab}, using that

$$(\partial_z G_{ab}) G_{bc}^{-1} = -\Omega_{ac}, \quad (33)$$

so that

$$\delta\Omega_{ab} \approx k^2 \sigma_v^2 \exp(7z/2) \begin{bmatrix} -28/375 & 28/375 \\ -4/75 & 4/75 \end{bmatrix}. \quad (34)$$

In order to see the role of decaying modes in the standard solutions of nonlinear PT, let's consider the usual second-order PT kernel (e.g. relevant for the calculation of the skewness and bispectrum). In our notation, the second-order kernel can be written as

$$\mathcal{F}_1^{(2)}(k_1, k_2) = \int_0^z ds \, g_{1b}(z-s) \, \gamma_{bcd} \exp(2s)(1, 1)_c(1, 1)_d. \quad (35)$$

where we assumed linear growing mode initial conditions. Since $g_{1b}\gamma_{bcd} = g_{11}\gamma_{121} + g_{12}\gamma_{222}$, and using Eq. (23) we have for the fastest growing contribution to 2nd-order PT

$$\begin{aligned}\mathcal{F}_1^{(2)}(k_1, k_2) &= \exp(2z)\left[\left(\frac{3}{5} + \frac{4}{35}\right)\alpha(k, k_1) + \left(\frac{2}{5} + \frac{4}{35}\right)\beta(k_1, k_2)\right] \\ &= \exp(2z)\left[\frac{5}{7}\alpha(k, k_1) + \frac{2}{7}\beta(k_1, k_2)\right]\end{aligned} \quad (36)$$

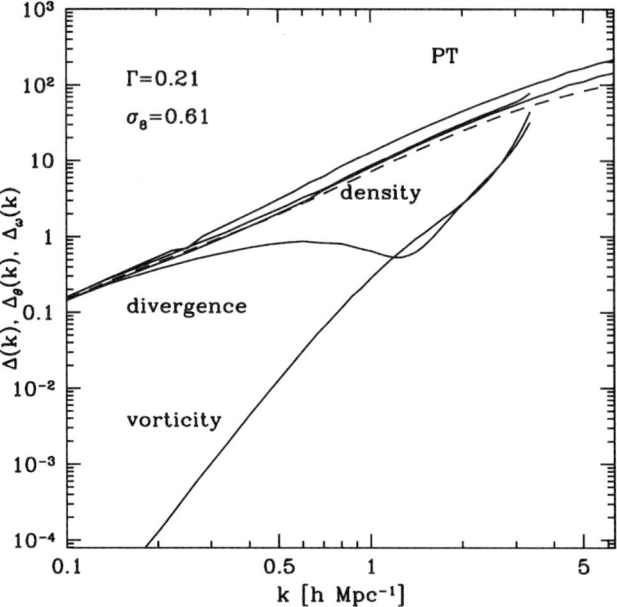

FIGURE 3. The power spectrum of the density, velocity divergence and vorticity as a function of scale. The two solid lines roughly parallel at high-k are the standard one-loop PT density power spectrum calculation (top) and the new one-loop approach (bottom). The dashed line denotes the prediction of the fitting formula and the solid line close to it the actual measurement in the N-body simulation. The two other solid lines denote the power spectrum of the velocity divergence and vorticity, as labeled.

It is crucial to note here that the 4/35 contributions come from *linearly decaying* modes; that is, after scattering, waves are not in linearly growing modes anymore, and this type of amplitude propagated into the present time contributes 4/35 to the amplitudes of second-order PT kernels. That means that if we are using the kernels to calculate one-loop corrections, which are important at intermediate k, one could use the approximation in Eq. (34) to improve the propagator in Eq. (35); in this case this corresponds to supressing the linearly decaying mode contribution to the propagator. As a result, the 2nd-order PT kernel at intermediate scales would look more like

$$\mathcal{F}_1^{(2)}(k_1, k_2) = \exp(2z)\left[\left(\frac{3}{5}\right)\alpha(k, k_1) + \left(\frac{2}{5}\right)\beta(k_1, k_2)\right]. \tag{37}$$

This means that a simple way of improving one-loop corrections in PT, is to use the kernels obtained by ignoring the linearly decaying modes contributions. This has the effect of incorporating higher-order loop corrections (those corresponding to propagator renormalization, although only approximately since we don't use the full one-loop propagator) in the usual formulation of PT.

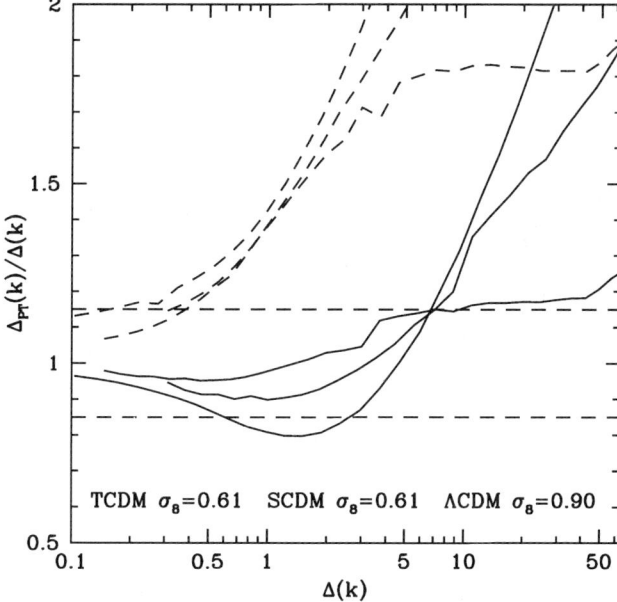

FIGURE 4. Ratio of predictions from one-loop PT (standard in dashed lines, new approach in solid lines) to the fitting formula for the nonlinear power spectrum. The three different curves are for three different models, as labeled.

If we supress linearly decaying mode contributions for the third-order kernel and use this to calculate one-loop corrections to the power spectrum, we find the results in FIGS. 3 and 4. The solid lines in FIG. 3 show the standard (top) and "improved" (bottom) calculations of the nonlinear power spectrum, whereas the dashed line shows the fitting formula for the nonlinear power spectrum. The three-remaining solid lines (which extend up to $k \approx 3$ h/Mpc) denote the measurement in N-body simulations of the density power spectrum, the velocity divergence power spectrum and the velocity vorticity power spectrum, as labeled. We see that the vorticity power spectrum is certainly negligible at large scales, and it does not become significant until scales of order $k \approx 2$ h/Mpc. At small scales, we find that the vorticity spectrum is roughly twice that of the divergence, as expected if the velocity field has equal power in all directions relative to k.

Note that the "improved" calculation is somewhat smaller than the standard one-loop calculation, as expected since the contribution from propagating linearly decaying modes has been suppressed. Overall the agreement with the N-body results is better. In FIG. 4 we show the ratio of our predictions for different models to the nonlinear fitting formula, the horizontal dashed lines show the expected accuracy of the latter. We see that the "improved" calculations (solid) stay within the nonlinear fitting formula accuracy up to $k \approx 5$ h/Mpc, whereas the standard one-loop calculation (dashed) overestimates the nonlinear power spectrum at scales smaller than the nonlinear scale, $k \approx 0.3$ h/Mpc. The improvement is thus quite significant, although we have included the effects of propagator renormalization in a crude way.

Unlike the velocity divergence, which can be calculated in one-loop PT in analogous fashion to the density power spectrum, understanding the vorticity power spectrum is considerably more complicated because vorticity is generated by shell-crossing, an effect which is neglected in the formulation of PT (see, e.g., [11]). However, we can understand approximately the scaling with redshift and scale from simple considerations. After shell-crossing, vorticity develops because what we see is the mass average of different streams, each with its own (irrotational) velocity field. Thus, vorticity can be thought as coming from the vorticity of the mass-weighted velocity field, i.e.,

$$w \sim f_v(\tau) \nabla \times [(1 + \delta)v], \qquad (38)$$

where $f_v(\tau)$ is the fraction of volume that undergoes shell-crossing at time τ. We can then write the vorticity power spectrum ($\langle w(k) \cdot w(k') \rangle \equiv P_w(k)\delta_D(k + k')$)

$$P_w(k) \sim f_v^2(\tau) \int \frac{|k \times q|}{q^4}[P_\theta(|k - q|)P_\delta(q) - P_x(|k - q|)P_x(q)]d^3q, \qquad (39)$$

where $P_\theta(k)$ is the velocity divergence power spectrum and $P_x(k)$ is the power spectrum of the density-velocity divergence cross-correlation. The simplest approximation would be to use linear PT (although it is unlikely to be valid for each flow at the scales of interest); however, since $P_\theta = P_\delta = P_x$ in linear PT, this contribution vanishes. Thus, the leading-order contribution to $P_w(k)$ comes from one-loop PT,

$$P_w(k) \sim f_v^2(\tau)k^2 \int \frac{d^3q}{q^2} a^6 |k - q|^{2n+3} q^n \sim a^6 f_v^2 k^{3n+6}, \qquad (40)$$

Thus, we expect a strong time and scale dependence for the vorticity power spectrum. The latter is in reasonable agreement with FIG. 3, the time dependence is more difficult to test due to the unknown dependence coming from $f_v^2(\tau)$.

4. CONCLUSIONS

We described a new approach to gravitational instability in large-scale structure, where the equations of motion are written and solved as in field theory in terms of Feynman diagrams. The basic objects of interest are the propagator (which propagates solutions forward in time), the vertex (which describes nonlinear interactions between waves) and a source with prescribed statistics which describes the effect of initial conditions. Loop corrections renormalize these quantities, in particular, decaying modes are supressed in the one-loop propagator compared to linear PT. We used this to construct the PT kernels and calculate "improved" loop corrections, these include effects beyond standard one-loop PT calculations, leading to better agreement with N-body simulations for the evolution of the power spectrum. We also consider the role of vorticity creation due to shell-crossing and show using N-body simulations that at small (virialized) scales the velocity field reaches equipartition, i.e. the vorticity power spectrum is about twice the divergence power spectrum. We also sketched a derivation of the time dependence and scaling of the vorticity power spectrum.

Our calculations are only a first attempt to include the effects of propagator renormalization, more work is needed to confirm that the results presented here are indeed robust to a more careful treatment. We have also neglected vertex renormalization. On the other hand, it seems that this approach can lead to useful insights into the nature of nonlinear corrections and perhaps give us a more accurate way to calculate clustering statistics in the transition to the nonlinear regime. Our results on the vorticity from numerical simulations suggest that there is a significant range of scales until the assumption of irrotational fluid breaks down. We hope that by using these techniques we can finally answer the question: "Why does PT work so well?"

ACKNOWLEDGMENTS

I thank Francis Bernardeau and Dmitry Pogosyan for useful discussions.

REFERENCES

1. GOROFF, M.H., B. GRINSTEIN, S.-J. REY & M.B. WISE. 1986. Coupling of modes of cosmological mass density fluctuations. Astrophys. J. **311**: 6–14.
2. FRY, J.N. 1984. The galaxy correlation hierarchy in perturbation theory. Astrophys. J. **279**: 499–510.
3. BERNARDEAU, F. 1992, The gravity-induced quasi-Gaussian correlation hierarchy. Astrophys. J. **392**: 1–14.
4. BERNARDEAU, F. 1994. The effects of smoothing on the statistical properties of large-scale cosmic fields. Astron. Astrophys. **291**: 697–712.
5. SCOCCIMARRO, R. & J.A. FRIEMAN. 1996. Loop corrections in non-linear cosmological perturbation theory. Ap. J. S. **105**: 37–73
6. SCOCCIMARRO, R. 1997. Cosmological perturbations: entering the non-linear regime. Astrophys. J. **487**: 1–17
7. SCOCCIMARRO, R., S. COLOMBI, J.N. FRY, J.A. FRIEMAN, E. HIVON & A. MELOTT. 1998. Non-linear evolution of the bispectrum of cosmological perturbations. Astrophys. J. **496**: 586–604.
8. FOSALBA, P. & E. GAZTAÑAGA. 1998. Cosmological perturbation theory and the spherical collapse model. I. Gaussian initial conditions. Mon. Not. R. Astr. Soc. **301**: 503–523.
9. GAZTAÑAGA, E. & P. FOSALBA. 1998. Cosmological perturbation theory and the spherical collapse model . II. Non-Gaussian initial conditions. Mon. Not. R. Astr. Soc. **301**: 524–534.
10. SCOCCIMARRO, R. 1998. Transients from initial conditions: a perturbative analysis. Mon. Not. R. Astr. Soc. **299**: 1097–1118.
11. PICHON, C. & F. BERNARDEAU. 1999. Astron. Astrophys. **343**: 663–681.

The Transition to Nonlinearity and New Constraints on Biasing

ROMAN JUSZKIEWICZ[a] AND ENRIQUE GAZTAÑAGA[b]

[a]*J. Kepler Astronomical Center, 65-265 Zielona Góra, Poland; Département de Physique Théorique, Université de Genéve, CH-1211 Genéve, Switzerland; and N. Copernicus Astronomical Center, 00-716 Warsaw, Poland*

[b]*INAOE, Astrofísica, Tonantzintla, Apdo Postal 216 y 51, Puebla 7200, Mexico; and Institut d'Estudis Espacials de Catalunya, IEEC/CSIC, Edf. Nexus-201 - c/ Gran Capitan 2-4, 08034 Barcelona, Spain*

ABSTRACT: We present two new dynamical tests of the biasing hypothesis. The first is based on the amplitude and the shape of the galaxy–galaxy correlation function, $\xi_g(r)$, where r is the separation of the galaxy pair. The second test uses the mean relative peculiar velocity for galaxy pairs, $v_{12}(r)$. This quantity is a measure of the rate of growth of clustering and it is related to the two-point correlation function for the matter density fluctuations, $\xi(r)$. Under the assumption that galaxies trace the mass ($\xi_g = \xi$), the expected relative velocity can be calculated directly from the observed galaxy clustering. The above assumption can be tested by confronting the expected v_{12} with direct measurements from velocity–distance surveys. Both our methods are checked against N-body experiments and then compared with the $\xi_g(r)$ and v_{12} estimated from the APM galaxy survey and the Mark III catalogue, respectively. Our results suggest that cosmological density parameter is low, $\Omega_m \approx 0.3$, and that the APM galaxies trace the mass at separations $r \gtrsim 5\ h^{-1}$, where h is the Hubble constant in units of 100 km s^{-1} Mpc. The present results agree with earlier studies, based on comparing higher-order correlations in the APM with weakly nonlinear perturbation theory. Both approaches constrain the linear bias factor to be within 20% of unity. If the existence of the feature we identified in the APM $\xi_g(r)$ — the inflection point near $\xi_g = 1$ — is confirmed by more accurate surveys, we may have discovered gravity's smoking gun: the long awaited "shoulder" in ξ, generated by gravitational dynamics and predicted by Gott and Rees 25 years ago.

KEYWORDS: Cosmology: theory, observation; Peculiar velocities: large-scale flows

1. INTRODUCTION

Recently, we have been working on two projects directly related to the transition to nonlinearity in gravitational clustering and present some of our results here. The first idea [submitted for publication in *Monthly Notices*] — to use the inflection

---

Address for correspondence: Roman Juszkiewicz, Departement de Physique Theorique, Universite de Geneve, 24, quai Ernest-Ansermet, CH-1211 Geneve 4, Switzerland.
roman@amorgos.unige.ch

point in the galaxy–galaxy correlation function to constrain biasing — is in a more mature state than the second, which uses measurements of the relative motions in pairs of galaxies. The results of our work on the second project are still far from completion but ripe enough to be presented and discussed.

1.1. The CDM Crisis and Biasing

The concept that galaxies may not be fair tracers of the mass distribution was introduced in the early eighties, in part in response to the observation that galaxies of different morphological types have different spatial distributions, hence they cannot all trace the mass (there are two excellent reviews on the subject: Strauss and Willick [1] and Hamilton [2]). However, there was also another reason: to "satisfy the theoretical desire for a flat universe" [3, p.391]. More precisely, biasing was introduced to reconcile the observations with the predictions of the Einstein–de Sitter cold dark matter (CDM) dominated model. At the time, it seemed that just a simple rescaling of the overall clustering amplitude by setting $\xi_g(r) = b^2 \xi(r)$, where $b \approx 2$ might do the job [3]. Very soon thereafter, however, it became clear that this is not enough: while the unbiased ($b = 1$) $\xi(r)$ had too large an amplitude at small r, the biased model did not have enough large-scale power to explain the observed bulk motions [4]. A similar conclusion could be drawn form comparison of the relative amplitude of clustering on large and small scales (e.g., [5]). The problem with the shape of $\xi(r)$ became explicit when measurements of $\xi_g(r)$ showed that the optically selected galaxies follow an almost perfect power law over nearly three orders of magnitude in separation. This result disagrees with N-body simulations. The standard ($\Omega_m = 1$) CDM model as well as its various modifications, including $\Omega_m < 1$ and a possible nonzero cosmological constant, fail to match the observed power law (see [6, Fig. 11-12], [7]; most of these problems were already diagnosed by Davis et al. [3]). Two alternative ways out of this impasse were recently discussed by Rees [8] and Peebles [9]. We believe that it will be helpful to discuss both approaches because their existence provides the motivation for our work.

1.2. Environmental Cosmology

A possible response to the CDM crisis is to build a model where simple phenomena, like the power-law behavior of ξ_g are much more complicated than they seem. In particular, one can explore the possibility that the emergence of large-scale structure is not driven by gravity alone but by "environmental cosmology" — a complex mixture of gravity, star formation and dissipative hydrodynamics [8]. A phenomenological formalism, appropriate for this approach was recently proposed by Dekel and Lahav [10]. According to the old, "linear biasing" prescription, at each smoothing scale, the galaxy- and the dark matter-density fields, δ_g and δ, are related by the linear function

$$\delta_g = b\, \delta, \qquad (1)$$

where b is a time- and scale-independent constant. In the new picture, the relationship between the two fields is neither linear nor deterministic. The biasing factor, here defined as the ratio of the rms values of the two fields, $b = \sigma_g/\sigma$ is allowed to depend on the smoothing scale as well as on the cosmological time. A convenient measure of the stochasticity in the relationship between the two fields is the cross-correlation coefficient,

$$R \equiv \frac{\langle \delta \delta_g \rangle}{\sigma_g \sigma}; \quad |R| \leq 1 \qquad (2)$$

The special case $R = 1$ describes the deterministic bias, while $R = 0$ corresponds to the "maximum stochasticity," or the case when the two fields are completely uncorrelated. The parameters R and b describe the biasing and stochasticity in the galaxy distribution relative to the spatial mass distribution. When referring to specific galaxy surveys, we will sometimes use subscripts, e.g., b_{IRAS} for the IRAS survey. These quantities should be distinguished from the bias and stochasticity measures for two classes of galaxies of different morphological types, e.g., for early (subscript e) and late (subscript l) galaxies: $b_{el} \equiv \sigma_e/\sigma_l$, and $R_{el} \equiv \langle \delta_e \delta_l \rangle (\sigma_e \sigma_l)^{-1}$.

The above parameters can be estimated or constrained by more or less direct measurements from galaxy redshift surveys, peculiar velocity data and other observations. They can be also studied in semi-analytic theoretical models [11], [12], in hydrodynamic simulations or in semi-analytic models combined with N-body simulations (e.g., [13] and [14] and references therein).

1.3. Constraints on Biasing

If biasing is indeed important and complicated, we should expect that b, b_{el}, R and R_{el} are all significantly different from unity and scale-dependent. As we show below, at sufficiently large scales, $r \gtrsim 10\, h^{-1}$ Mpc, there is actually evidence to the contrary: the admissible deviations of b and R are small and always comparable to the accuracy of the measurements.

Skewness and higher moments. The strongest constraints on large (weakly nonlinear) scales come from the measurements of the two-, three-, and four-point connected moments of the density field in the APM catalogue. These constraints are obtained as follows. One assumes that galaxies trace the mass and the large-scale structure we observe today grew out of small-amplitude, Gaussian density fluctuations in an expanding, self-gravitating nonrelativistic gas. If our assumption is correct, by now nonlinear gravitational instability would have driven the distribution away from gaussianity, generating skewness and higher connected moments, which can be calculated analytically [15]–[17]. The assumptions about initial gaussianity and lack of biasing can then be tested by comparing the analytical predictions with observations. The predictions are in good agreement with the data from the APM [18]–[20] as well as the IRAS PSCz catalogue [21] and suggest that $b(r)$ is within 20% of unity for $r \gtrsim 10\, h^{-1}$Mpc, or at linear scales (where the clustering amplitude is less than unity).

Redshift distortions. Some of the most radical claims that b_{el} can be as large as 1.5 to 2 are based on comparisons of the estimates of the strength of clustering in the 1.2 Jy IRAS catalogue with optical redshift surveys ([1] and references therein). Indeed, those earlier studies, summarized by Hamilton [2], gave a redshift distortion parameter $\beta_{IRAS} = \Omega_m^{0.6}/b_{IRAS} = 0.77 \pm 0.22$, while the average and the standard deviation of the same parameter, obtained from optical surveys is, according to the same compilation, $\beta_{optical} = 0.52 \pm 0.26$, implying a relative bias $b_{optical}/b_{IRAS} = b_{el} \approx 1.5$. The most recent analysis [22] of the larger PSCz sample [23] gives $\beta_{IRAS} = 0.41 \pm 0.13$, consistent with $b_{el} = 1$. Moreover, Hamilton et al. also conclude that their results are consistent with $R_{IRAS}(r) \approx 1$ at $r \gtrsim 10\ h^{-1}$ Mpc. More recently, Hamilton and Tegmark [22] have studied the large as well as small-scale clustering in real space, reconstructed from the redshift space data of the PSCz survey and found evidence for scale-dependent bias; however, the effect is significant at small, strongly-nonlinear scales only.

Large-scale flows. Recent measurements of the mean relative pairwise velocity of galaxies allow an independent estimate of Ω_m and the biasing parameter. These results are consistent with $\Omega_m \approx 0.3$ and $R \approx b \approx 1$, $b_{el} \approx R_{el} \approx 1$ at separations $r \gtrsim 5\ h^{-1}$ Mpc [23].

Weak lensing. The values of b and Ω_m, derived from large-scale flows are consistent with recent measurements of cosmic shear correlations based on the VIRMOS deep imaging survey [24]. On much smaller scales, $r \approx 1\ h^{-1}$ Mpc, weak lensing data from the Sloan Survey give $\Omega_m R/b \approx 0.3$ [25], again in agreement with the estimates from the pairwise motions, although unlike the relative velocity approach, the Sloan results suffer from degeneracies between Ω_m and bias.

Qualitative arguments. Qualitatively, strong biasing effects would be difficult to reconcile with the well-known fact that the L_* galaxies, dwarf galaxies and IRAS galaxies have strikingly similar distributions, all avoiding the voids [26, pp. 638–642]. Another empirical argument against biasing is provided by the universal character of the observed Tully–Fisher and fundamental plane relations (see, e.g., [9], [27]).

Simulations. All simulations of the galaxy formation process, either of hydrodynamic or semi-analytic variety predict morphological segregation as well as b and R parameters dependent on scale and time. Since these models are at a relatively early stage of their development, the details differ from model to model on small scales and at high redshifts, i.e. exactly where their b and R parameters are significantly different from unity and where clustering is strongly nonlinear. However, most of these numerical experiments broadly agree that with decreasing redshift and increasing scales, b and R approach unity (see, e.g., [28, Fig. 18] and references therein).

In the end, it may turn out that to explain the observed structure all we need is just the plain gravitational instability theory, leaving complex non-gravitational physics on scales below a Megaparsec to cosmogony, directly involving star formation. We discuss this possibility below.

1.4. What You Get Is What You See

An obvious alternative to environmental cosmology was recently discussed by Peebles [9], who pointed out that "as Kuhn has taught us, complex interpretations of simple phenomena have been known to be precursors of paradigm shifts" and perhaps after fifteen years of attempts to salvage the CDM model with biasing, it is time to abandon this approach, as well as the biasing hypothesis itself as "another phlogiston" (all quotations are from [9]). Instead, one can explore a simpler option, that galaxies trace the mass distribution, or

$$\xi_g = \xi \quad \text{and} \quad R = b = 1 \tag{3}$$

at least for local (low redshift), optically selected galaxies with a broad magnitude sampling. This approach rests on the idea that no matter how or where galaxies form, they must eventually fall into the dominant gravitational wells and therefore trace the underlying mass distribution (see [29], [30]). Our purpose here is to test this idea, using measurements of relative velocities of pairs of galaxies and the shape of the two-point correlation function.

1.5. Outline of the Paper

In this paper we propose two new tests of the biasing hypothesis, which involve two measures of clustering. The first is the two-point correlation function of mass density fluctuations, ξ. The second is a measure of the rate of gravitational clustering — the relative velocity of particle pairs, v_{12}. We describe our theoretical model in the next section. Our analytic formulas used to test the biasing hypothesis are checked against N-body simulations in Section 3. In Section 4 we apply our tests to the APM galaxy survey. Finally, in Section 5 we summarize and discuss our results.

2. THEORY

2.1. The Relative Velocity

The relative pairwise velocity v_{12} was introduced in the context of the BBGKY theory [31], describing the dynamical evolution of a collection of particles interacting through gravity. In this discrete picture, $\vec{v}_{12}$ is defined as the mean value of the peculiar velocity difference of a particle pair at separation $\vec{r}$ [29, Eq. 71.4]. In the fluid limit, its analogue is the pair-density weighted relative velocity [32],

$$\vec{v}_{12}(r) = \frac{\langle (\vec{v}_1 - \vec{v}_2)(1 + \delta_1)(1 + \delta_2) \rangle}{1 + \xi(r)} \tag{4}$$

where $\vec{v}_A$ and $\delta_A = \rho_A / \langle \rho \rangle - 1$ are the peculiar velocity and fractional density contrast of matter at a point $A = 1, 2, \ldots$. The brackets $\langle \ldots \rangle$ denote ensemble averages for pairs of points at a fixed separation $r = |\vec{r}_1 - \vec{r}_2|$, while $\xi(r) = \langle \delta_1 \delta_2 \rangle$. In gravitational instability theory, the magnitude of $\vec{v}_{12}(r)$ is related to $\xi(r)$ through the pair conservation equation [29, Eq. 71.6].

Recently, Juszkiewicz et al. [34] have proposed an approximate solution of the pair conservation equation. Using Eulerian perturbation theory, they solved the equation for $v_{12}(r)$ to second order in ξ. They also proposed an interpolation between their large-r perturbative limit, and the well-known small separation limit — the stable clustering solution, $v_{12}(r) = -Hr$, where H is the Hubble constant [29, Sect. 71]. The resulting ansatz is given by

$$v_{12}(r) = \frac{2}{3} H f r \bar{\bar{\xi}}(r)[1 + \alpha \bar{\bar{\xi}}(r)] \tag{5}$$

$$\bar{\xi}(r) = (3/r^3) \int_0^r \xi(x) x^2 dx \equiv \bar{\bar{\xi}}(r)[1 + \xi(r)]. \tag{6}$$

Here α is a parameter, determined by the logarithmic slope of $\xi(r)$, while $f = d \ln D / d \ln a$, with $D(a)$ being the standard linear growing mode solution and a — the cosmological expansion factor [29, Sect. 11]. The solution (5) assumes Gaussian initial conditions, and the dynamics of clustering is assumed to be dominated by the gravity of inhomogeneities in a pressureless fluid of nonrelativistic particles. For all such models, including those with a nonzero cosmological constant, $f \approx \Omega_m^{0.6}$ [26, Sect. 13]. If the correlation function is given by a pure power law, $\xi \propto r^{-\gamma}$, where $0 < \gamma < 3$, the parameter α is given by

$$\alpha \approx 1.2 - 0.65\gamma \tag{7}$$

This formalism can be generalized for a nonpower law $\xi(r)$ by replacing γ in Eq. (7) with an effective slope,

$$\gamma_o \equiv -d \ln \xi / d \ln r \big|_{r_o}. \tag{8}$$

evaluated at separation $r = r_o$, defined by the condition

$$\xi(r_o) = 1. \tag{9}$$

Juszkiewicz et al. [34] tested Eqs. (5)–(9) against high resolution N-body simulations, provided by the Virgo consortium [7]. They found an excellent agreement between the streaming velocity, predicted by their ansatz and the $v_{12}(r)$, measured from the simulations in the entire dynamical range, $0.1 < \xi < 10^3$. However, the N-body experiments they used were confined to four different CDM-like models, considered by Jenkins et al. [7]. As we have already pointed out in the INTRODUCTION, models of this kind fail to reproduce the observed $\xi_g(r)$ unless one resorts to a highly contrived, scale- and time-dependent biasing function [6], [7], [9]. One of our objectives here is to test the validity Juszkiewicz et al. [34] ansatz for $v_{12}(r)$ against a new set of N-body simulations, which differ significantly from those originally considered by Juszkiewicz et al. [34]. Here we use simulations with a more realistic $\xi(r)$, inferred by Baugh and Gaztañaga [35] from the measurements of galaxy–galaxy correlations in the APM survey (see the description below; from now on, we will refer to these numerical experiments and their initial conditions as APM-like).

2.2. The Inflection Point

In the gravitational instability theory, newly forming mass clumps are generally expected to collapse before relaxing to virial equilibrium. If this were so, $|v_{12}(r)|$ would have to be larger than the Hubble velocity Hr to make $v_{12}(r) + Hr$ negative. As a consequence, the slope of ξ,

$$d \ln \xi(r)/d \ln r = -\gamma(r), \qquad (10)$$

must increase at separations where $\xi(r, t) \approx 1$. This effect was recognized long ago by Gott and Rees [36]. When the expected "shoulder" was not found in the observed galaxy–galaxy correlation function, Davis and Peebles [37] introduced the so-called previrialization conjecture as a way of reducing the size of the jump in $\gamma(r)$ (the conjecture involves nonradial motions within the collapsing clump; see the discussion in [29, Sect.71] and [26, pp. 535–541]; see also [38]–[40]). Later observational work showed a clear break in the shape of ξ for several redshift and angular catalogues, which was early evidence for the linear to non-linear transition, pointed out by Guzzo and collaborators (see the review by Guzzo [41]).

The arguments, raised by Peebles and Davis [37] were qualitative rather than quantitative. A quarter of a century later the precision of N-body simulations as well as the quality of the observational data have improved dramatically enough to justify a reexamination of the problem. The actual shape of the correlation function near $\xi = 1$ can be investigated with high resolution N-body simulations like those run by the Virgo Consortium [7]. In all four of the Virgo models, the slope of $\xi(r)$ exhibits a striking feature. Instead of a shoulder, or a simple discontinuity in $\gamma(r)$, however, $\xi(r)$ has an inflection point,

$$d^2\xi(r)/dr^2 = 0, \qquad (11)$$

which occurs at a uniquely defined separation $r = r_*$. At this separation, the logarithmic slope of ξ reaches a local maximum, $d \ln\xi/d \ln r = -\gamma_*$. In all four of the models Juszkiewicz et al. [34] investigated, the inflection point indeed appears near the transition $\xi = 1$, as expected by the earlier speculations, involving the "shoulder" in ξ. Namely, r_* is almost identical with the scale of nonlinearity:

$$r_* \approx r_o. \qquad (12)$$

More precisely, a comparison of Figure 1 in Juszkiewicz et al. [34] with Figure 8 in Jenkins et al. [7] gives

$$|r_o - r_*| < 0.1 r_o \qquad (13)$$

for all four considered models. Moreover, for all models, studied by Juszkiewicz et al. [34], the $-\gamma$ vs. r dependence can be described as an S-shaped curve, with a maximum at $r = r_* \approx r_o$, and a minimum at a smaller separation. For $r \geq r_*$, the nonlinear slope (measured from Virgo simulations) follows the linear $\gamma(r)$, determined by the initial conditions. The separation r_* is therefore also the branching point between the linear and nonlinear $\gamma(r)$ curves, which are identical for $r > r_*$ (actually, they differ a little; the small differences between the two curves can be entirely attributed to sampling errors in the N-body experiments; see [34, Fig. 1]). This property can be used to identify r_* in noisy simulations, and/or observations, when the noise in the measured $\gamma(r)$ curve does not allow a direct determination of r_* as the position of the maximum in $-\gamma(r)$.

If relation (12) is indeed a general property of gravitational clustering, it can be used as a test of biasing as follows. Suppose the biasing factor is a scale-independent constant, significantly greater than unity: $b \gg 1$. Then $\xi_g \gg \xi$ and relation (12) will break down. For a power-law correlation function, $\xi_g(r) = (r_o/r)^\gamma = b^2 \xi(r)$, and instead of Eq. (12) we will have

$$r_* \approx r_o b^{-2/\gamma}. \qquad (14)$$

Since the observed slope is $\gamma \approx 1.8$, for $b = 2$, the shoulder in the correlation function should appear at a separation smaller than a half of the r_o! The above argument can be generalized to a broader class of models, allowing scale-dependent bias provided $b(r)$ is a smooth monotonic function and $b(r_o)$ is significantly greater than unity. Whenever these conditions apply, strong biasing always implies $r_o \gg r_*$. Hence, the comparison of these two scales determined from observations can be used as a diagnostic for biasing.

3. N-BODY SIMULATIONS

3.1. Initial Conditions

In this section, we compare our ansatz for $v_{12}(r)$ with the results from P^{3}M simulations. We use APM-like models for the initial shape of $P(k)$ in Baugh and Gaztañaga [35]. The box size is 600 h^{-1} Mpc (or 300 h^{-1} Mpc) with 200^3 (or 100^3) dark matter particles. The APM-like models have Gaussian initial conditions with a power spectrum inferred from correlations in the APM survey, following the procedure, introduced by Baugh and Gaztañaga [35]. The power spectrum of the density fluctuations is related to $\xi(r)$ through the usual Fourier transform,

$$P(k) = \int \xi(r) \exp(i\mathbf{k} \cdot \mathbf{r}) \, d^3r. \qquad (15)$$

Following the prescription of Baugh and Gaztañaga, we assume an initial power spectrum of the form

$$P(k) = \frac{Ak}{(1 + (k/0.05h \text{ Mpc}^{-1})^2)^{1.6}} \qquad (16)$$

The linear spectrum, given above is designed to evolve into a nonlinear one, matching the APM measurements of $P(k)$ under the assumption of no bias. The normalization constant A is directly related to the degree of inhomogeneity of the mass distribution at the end of the simulation, parametrized by σ_8 — the rms mass density contrast, measured in spheres of a radius of 8 h^{-1} Mpc. Following Baugh and Gaztañaga, we choose the constant A in order to ensure that at the end of the simulation, $\sigma_8 = 0.85$. The quality of the agreement of the evolved $\xi(r)$ with the APM data at small separations, where $\xi \gg 1$, depends on the assumed value of Ω_m. However, as we show below, for $r \geq 2\ h^{-1}$ Mpc and $\xi < 3$, this effect is negligible (compare FIGURES 1 and 3). This range of separations and clustering amplitudes brackets from both sides the region, on which we focus on here: the transition between $\xi < 1$ and $\xi > 1$. We use simulations with $\Lambda = 0$ and two different values of the density parameter: $\Omega_m = 1$ and 0.3.

3.2. Estimators

Two different estimators for $\xi(r)$ and v_{12} are used. To construct the first estimator, we cut the simulation box into cubical pixels of size Δ_1, placed on a regular lattice. The density fluctuation amplitude at the ith pixel is

$$\delta_i = \frac{N_i}{\langle N \rangle} - 1, \tag{17}$$

where N_i is the particle count in that pixel. The estimator for the two-point function is then:

$$\hat{\xi}^{(1)}(r) = \frac{1}{N_r} \sum_{i,j} \delta_i \delta_j W_{ij}(r), \tag{18}$$

where

$$N_r = \sum_{i,j} W_{ij}(r), \tag{18}$$

is the number of pairs of pixels at separation r in the sampled region, and the window function $W_{ij}(r) = 1$ if pixels i and j are separated by $|\vec{r}_i - \vec{r}_j| = r \pm \Delta r$, and 0 otherwise. For the pairwise velocity we define as $\hat{v}_i$ the average velocity in the ith pixel at position $\vec{r}_i$. We can then use an equivalent expression for the first estimator of v_{12}:

$$\hat{v}^{(1)}(r) = \sum_{i,j} (\vec{v}_i - \vec{v}_j) \cdot \hat{r}_{ij}\, U_{ij}(r) \tag{20}$$

where

$$\hat{r}_{ij} \equiv \frac{\vec{r}_i - \vec{r}_j}{|\vec{r}_i - \vec{r}_j|} \tag{21}$$

is a unit vector separating pixels i and j and the sum is over all pairs of pixels, while

$$U_{ij} \equiv \frac{(1 + \delta_i)(1 + \delta_j) W_{ij}(r)}{N_r[1 + \xi(r)]} \tag{22}$$

The second estimator uses all of the particle pairs rather than the counts in pixel pairs:

$$\hat{\xi}^{(2)}(r) = \frac{1}{\langle n \rangle V_r} \sum_{i>j} W_{ij}'(r) - 1 \tag{23}$$

where now the window function $W_{ij}'(r) = 1$ if particles i and j are separated by $|\vec{r}_i - \vec{r}_j| = r \pm \Delta r$, $\langle n \rangle$ is the mean particle density and V_r is the volume of the spherical shell of radius r and thickness Δr. Typically, $\langle n \rangle V_r$ is estimated from a random catalogue of particles, drawn from the same sample, because the effective volume, containing the pairs separated by a distance $r \pm \Delta r$ might depend on the geometry. However, our simulations use a large and periodic box, with no boundaries and high densities (not affected by shot-noise at the scales of interest). So the above expression gives a good and quick estimator. The corresponding estimator for the pairwise velocity is

$$\hat{v}^{(2)}(r) = \frac{\sum_{i>j}(\vec{v}_i - \vec{v}_j) \cdot \hat{r}_{ij} W_{ij}'(r)}{\sum_{i>j} W_{ij}'(r)} \tag{24}$$

where v_i and v_j are now individual particle velocities. Note that this estimator does not depend on the effective volume of the shell, V_r.

Both estimators agree reasonably well in our simulations. The first set is more useful (faster to run) for large separations, as we can reduce the resolution of the lattice and have a relatively small number of pixels. The second set is more adequate (faster to run) for the small separations.

3.3. The Correlation Function

The evolved, nonlinear correlation functions, measured from simulations are shown in FIGURE 1 (*top panel*). The full squares correspond to the $\Omega_m = 1$ model, while the open squares represent $\Omega_m = 0.3$. For comparison, we show the linear correlation function (dashed line), scaled from some "initial" redshift z_o to the present ($z = 0$), following the linear theory expression for the Einstein–de Sitter model, $\xi(r, 0) = \xi(r, z_o)(1 + z_o)^2$. Nonlinear effects are more pronounced in the low density model. This happens because of the well-known suppression of linear growth, which occurs at late times ($z < 1/\Omega_m$) in low density models and leads to the enhanced clustering on small scales relative to large scales.

Note however, that although the correlation functions differ significantly in amplitude at separations $r < 2\ h^{-1}$ Mpc, their slopes $\gamma(r)$ are almost indistinguishable (FIG. 1, *bottom panel*).

The particle resolution (the Nyquist wavelength $\propto N^{-1/3}$) of the simulations used here is significantly lower than the resolution of Virgo simulations, and the noise in the measured $\xi(r)$ is further amplified by differentiating over r. As a result, determining the position of the inflection point r_* directly from the $\gamma(r)$ curve alone is diffi-

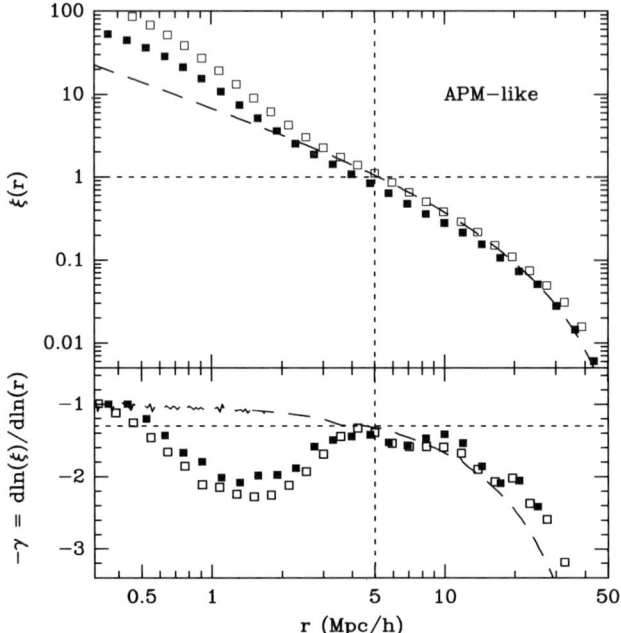

FIGURE 1. The *top panel* shows the linear correlation function (dashed line) and the measured non-linear ξ(r), obtained from the APM-like simulations with density parameters $\Omega_m = 0.3$ (*open squares*) and $\Omega_m = 1.0$ (*full squares*). The bottom panel shows the corresponding logarithmic slope, $-\gamma(r) = d \ln \xi / d \ln r$ for each of the three curves from the top panel. The vertical dotted line shows the separation r_o, defined by the condition $\xi(r_o) = 1$ (top) and the separation r_*, at which the nonlinear $\gamma(r)$ curve crosses the linear one (bottom).

cult. To overcome this problem, we identify r_* by comparing the linear and nonlinear $\gamma(r)$ curves. Taking r_* to be the separation at which the nonlinear slope drops below the linear slope in FIGURE 3, as described earlier, we get

$$r_* \simeq 5\, h^{-1}\, \text{Mpc} \approx r_o, \qquad (25)$$

in excellent agreement with Eq. (13). Hence, the equality between r_* and r_o can probably be considered as a generic outcome of gravitational dynamical evolution in a model where galaxies trace the mass and the initial slope, $d \ln \xi / d \ln r$, is a smooth decreasing function of the separation r. Such a picture is also known as hierarchical clustering; an obvious additional condition to make sure that small scale clumps collapse before the large-scale ones, is $\gamma > 0$, see e.g. [29].

3.4. The Relative Velocity

In this section we describe N-body tests of the Juszkiewicz *et al.* [34] model for the relative motions in pairs of galaxies. We consider two models with APM-like ini-

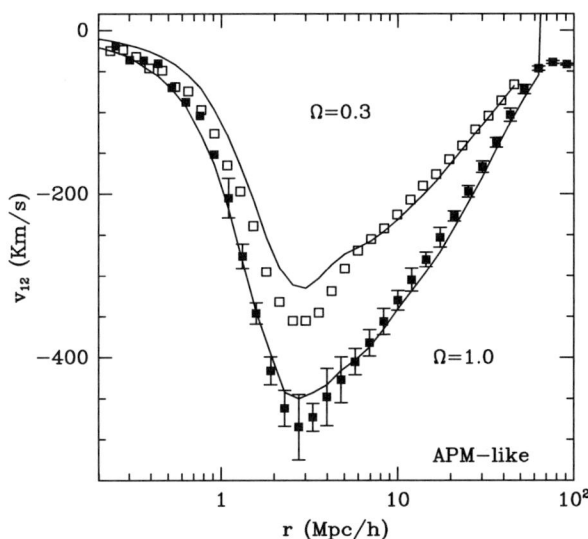

FIGURE 2. The mean pairwise velocity $v_{12}(r)$, measured from two sets of APM-like simulations with $\Omega_m = 0.3$ (*open squares*) and $\Omega_m = 1.0$ (*full squares*), are compared with with an approximate analytical solution of the pair conservation equation (Eq. (5), *continuous lines*).

tial spectra: an Einstein–de Sitter model and an open model with $\Omega_m = 0.3$. For both models, the theoretical predictions for the mean pairwise velocity, based on Eq. (5), are plotted in FIGURE 2 as continuous lines. These predictions can be compared with N-body measurements, shown as full squares for the $\Omega_m = 1$ model and as open squares for $\Omega_m = 0.3$. The agreement between the theoretical and experimental $v_{12}(r)$ curves shows that our ansatz provides a good approximation of the N-body results in the entire dynamical range probed for both models. The mean and errors in the $\Omega_m = 1$ simulations come from the mean and dispersion, obtained from five independent realizations of the APM-like model. For the open model ($\Omega_m = 0.3$, *open squares*) we used only one realization, but the expected sampling variance is expected to be the same. Indeed, the initial $P(k)$ is identical in both cases. Moreover, the longwave tails of the *final* power spectra (which determine the size of the sampling error bars) are also identical because they are not affected by the nonlinear evolution.

4. COMPARISON WITH OBSERVATIONS

4.1. The Correlation Function

The measurements of $\xi_g(r)$, and $\gamma(r) \equiv -d \ln \xi_g / d \ln r$, obtained from the angular correlations of galaxy pairs in the APM catalogue [42], are plotted in FIGURE 3. The top panel shows the two-point function (points with error bars), and the linear theory

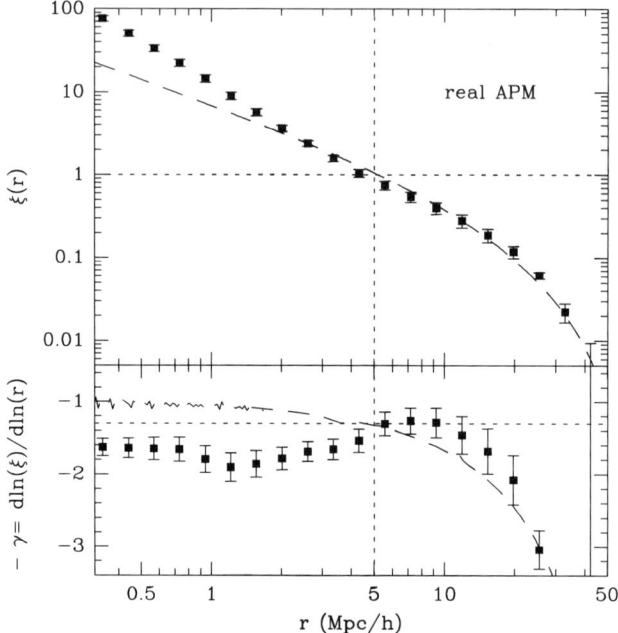

FIGURE 3. The spatial correlation function of APM galaxies (*top panel, symbols with errorbars*), compared to the linear theory APM-like model, described earlier (*top panel, dashed line*). The bottom panel shows the corresponding logarithmic slope, $\gamma(r)$. The intersection of the two perpendicular dotted lines marks the points where $\xi \simeq 1$ (*top*) and where the nonlinear slope crosses the linear one (*bottom*).

curve, described in Section 3.3 (*dashed line*). The intersection of the two perpendicular dotted lines marks the point $(\xi_g, r) = (1, r_o)$. The bottom panel of FIGURE 3 shows the APM $\gamma(r)$ as a function of the pair separation r. Note the remarkable similarity between the empirical data and the characteristic peak in the $\gamma(r)$ found in the simulations (FIG. 1). The intersection of the two mutually perpendicular, *dotted lines* in the bottom panel of FIGURE 3 marks the crossing between the linear model for $\gamma(r)$ (*dashed line*) and the nonlinear $\gamma(r)$ curve, determined from the APM catalogue. The crossing occurs at the separation $r \simeq 5\ h^{-1}$ Mpc, and to first approximation this scale could be identified with r_*. However, a closer inspection of our FIGURE 3 suggests that, given the error bars, the actual position the peak could be shifted to the right, to a somewhat larger separation. Taking into account the error bars in FIGURE 3 as well as the uncertainties in the assumed linear theory slope, we obtain

$$r_* \simeq (6 \pm 1)\ h^{-1}\ \text{Mpc}, \tag{26}$$

and

$$r_o \simeq (5 \pm 1)\ h^{-1}\ \text{Mpc}. \tag{27}$$

The slope at $r = r_*$ is $\gamma_* \simeq -1.4$. If we assume the linear bias model, the relation $r_* \approx r_o b^{-2/\gamma_*}$ gives

$$b = 1.11 \pm 0.22 \qquad (28)$$

at one-sigma statistical significance level.

4.2. The Relative Velocity

We will now apply the second of the two proposed tests of biasing: the relative velocity test. We will compare the mean pairwise velocity, predicted by assuming that the APM galaxies trace the mass with the pairwise velocity, measured directly from a peculiar velocity — distance survey.

FIGURE 4 shows $v_{12}(r)$ curves, predicted by Eq. (5) (*continuous lines*) for three different values of Ω_m, from bottom to top $\Omega_m = 1.0, 0.3, 0.1$, respectively. To calculate $v_{12}(r)$, we have used $\xi(r)$, estimated from the APM survey under the assumptions $\xi_g = \xi$ and $R = 1$.

Before making the comparison, we must overcome the following problem. The survey has a significant depth, with the mean redshift of $z \simeq 0.15$ while the measured $v_{12}(r)$ corresponds to the present time ($z = 0$). To evolve $\xi(r, z)$ from $z \simeq 0.15$ to $z = 0$, we need to make some additional model assumptions. Gaztañaga [6] has shown that for this redshift range, the uncertainties in the details of dynamical evolution of ξ are small. In particular, choosing an incorrect value for Ω_m can affect ξ at most at the several percent level (for Ω_m ranging from 1 to 0). Adding this to other possible sources of errors, such as uncertainties regarding the redshift evolution of the galaxy number density and sampling and selection fluctuations, Gaztañaga [6] estimates that the resulting relative uncertainty in the amplitude of ξ_g is $\lesssim 20\%$. According to his analysis, the present ($z = 0$) amplitude of the rms fluctuation of the APM galaxy counts, measured in spheres of radius of $8\ h^{-1}$ Mpc, is $1.1 \lesssim \sigma_8^{APM} \lesssim 0.9$.

To be conservative, for each value of Ω_m, we plot the predicted $v_{12}(r)$ curves for two values of σ_8, differing by 20%. The resulting prediction for each value of Ω_m is therefore an area rather than a single $v_{12}(r)$ curve (see FIG. 4). The lower boundary of each shaded area assumes $\sigma_8 = 1.1$ while the upper boundary was calculated by assuming $\sigma_8 = 0.9$. A direct measurement at $r = 10\ h^{-1}$ from the Mark III galaxy peculiar velocity survey [43] gives [23]

$$v_{12} = -280 \pm 60 \text{ km/s}. \qquad (29)$$

It is reassuring that this value, plotted in FIGURE 4, overlaps with the shaded area, corresponding to $\Omega_m = 0.3$, because it agrees with ranges for Ω_m and σ_8 obtained by Juszkiewicz *et al.* from the analysis of the Mark III survey alone. Their one-sigma constraints are $\Omega_m = 0.35^{+0.35}_{-0.25}$ and $\sigma_8 \geq 0.7$, and the analysis assumes that the correlation function for the mass is well approximated by a pure power law with $\gamma = 1.75$. From the agreement between the predicted and observed value of the mean pairwise velocity, we conclude that the Mark III and APM data, taken together, are consistent with the hypothesis that the APM galaxies trace the mass, $b \approx R \approx 1$, while the density parameter is low, $\Omega_m \approx 0.3$.

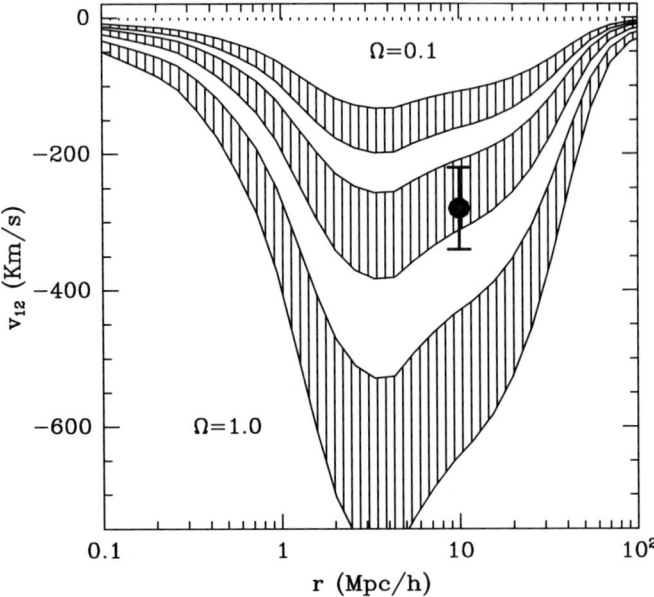

FIGURE 4. Predictions for the mean pairwise velocity v_{12}, based on the assumption that the APM galaxies trace the mass. The shaded regions correspond to 20% uncertainties in the strength of clustering. We consider three values of $\Omega_m = 1.0, 0.3, 0.1$, top to bottom (as labeled). The point with error bars corresponds to a direct measurement from the Mark III survey [23].

4.3. How Can Biasing Affect $v_{12}(r)$

Apart from leading to predictions which are confirmed observationally, the "what you get is what you see" hypothesis has another important advantage: simplicity. Once $\xi(r)$ is estimated from observations, a family of $v_{12}(r, \Omega_m)$ curves can be calculated directly from Eq. (5) for any given range of values of the density parameter. This simplicity will immediately go away if we allow scale-dependent, stochastic and nonlinear biasing. A frank answer to the question posed in the heading of this subsection would have to be "Only God (of biasing) knows." Predicting $v_{12}(r)$ without resorting to massive numerical simulations would be simply impossible. We can get the idea of what is in store by considering only the leading order term in the perturbative expansion for $v_{12}(r)$ at large separations [44],

$$v_{12}(r) = -\frac{2}{3} H \, r f(\Omega_m) \bar{\xi}_{g\rho}(r), \tag{30}$$

where $\bar{\xi}_{g\rho}(r) = (3/r^3)\int_0^r \xi_{g\rho}(x) x^2 dx$ is the galaxy-mass cross-correlation function,

$$\xi_{g\rho}(r) = \langle \delta(0)\delta_g(\vec{r}) \rangle, \tag{31}$$

averaged over a sphere of radius r. The function $\xi_{g\rho}$ describes the cross-correlations between the mass density and the density of galaxies in the velocity field survey, which in the case considered here would be the Mark III catalogue.

To make progress in our analysis, we will now generalize the definition of the stochasticity parameter introduced as a normalized cross-correlation of two random fields, δ and δ_g, measured at the same position in space. Instead, we will consider a cross-correlation of the same two fields measured at two different positions in space, separated by distance r. Our old Eq. (2) is replaced by

$$R(r) = \frac{\xi_{g\rho}(r)}{\sqrt{\xi(r)\xi_g(r)}}. \tag{32}$$

Let us make another simplifying assumption, that $b^2(r) \equiv \xi_g(r)/\xi(r)$, as well as R, are separation-independent. Equation (30) becomes

$$v_{12}(r) = -\frac{2}{3} f(\Omega_m) H \, r \, Rb\bar{\xi}(r). \tag{33}$$

The expected relative pairwise velocity can now be related to the APM data by substituting

$$v_{12}(r) = \bar{\xi}(r) = \frac{3}{b_A^2 r^3} \int_0^r \xi_g(x) x^2 dx. \tag{34}$$

where $b_A^2 = \xi_g(r)/\xi(r)$ and ξ_g is the APM galaxy correlation function. In case of trouble in predicting the correct $v_{12}(r)$ curve, we now have three essentially free parameters which can be readjusted. This is only the tip of the iceberg, as we have ignored nonlinear dynamics as well as the scale-dependence of b and R.

For the linear bias model, the predictions are in clear conflict with the data unless b is close to unity. After setting $R = 1$, we get $v_{12}(r) \propto b$. Then, if the biasing factor for spiral galaxies is, as usually assumed $b \approx 1$, our predictions for $v_{12}(r)$ in the linear regime ($r \gtrsim 10 \, h^{-1}$ Mpc) should be similar to the unbiased predictions already plotted in FIGURE 4. If the biasing factor for the ellipticals is significantly different, say, $b \approx 2$, the elliptical subsample of the Mark III data should give estimates of v_{12} which differ from the estimates from the spiral sample by the same factor of two. Meanwhile the estimates from the appropriate subsamples in the real data are indistinguishable [23]. Hence, just as in case of the shape of the APM correlation function, considered above, the deterministic linear biasing model is inconsistent with observations.

We can summarize the last two subsections as follows. The prediction for $v_{12}(r)$, based on the assumption that the APM galaxies trace the mass passes our test as it agrees with the velocity, estimated from the Mark III data. The simplest prescription of biasing fails the test. More complicated prescriptions can probably be made to pass, which is not surprising, given the number of free parameters.

5. SUMMARY AND DISCUSSION

A quarter of a century ago, Gott and Rees predicted that gravity should leave its mark on the shape of the galaxy autocorrelation function: a "shoulder," or steepening of the slope of the correlation function should appear near the separation where ξ passes through unity. At the time, biasing was unheard of, and Gott and Rees [36] assumed $\xi = \xi_g$. Recently, in another context, Juszkiewicz et al. [34] have studied the $\xi = 1$ boundary in the evolution of the mass correlation function, using results from Virgo simulations. They found that the "shoulder" is actually an inflection point, occurring at a well-defined separation r_*. In all four CDM-like models they studied, the nonlinear transition looked strikingly similar: the inflection occurred at almost the same separation as that of the nonlinear transition: $r_* \approx r_o$, where r_o corresponds to $\xi = 1$. Here we have tested the degree of universality of their results by widening the range of models considered. Our additional objective was to study the range of validity of an approximate solution of the pair conservation equation, proposed by Juszkiewicz et al. [34] to study the nonlinear evolution of the relative velocity of particle pairs at a fixed separation, $v_{12}(r)$. We used N-body simulations, with APM-like initial conditions, with two different values of the density parameter: $\Omega_m = 1$ and 0.3. The APM-like initial power spectra differ significantly from all of the CDM-like spectra, considered earlier by Juszkiewicz et al. [34]. Moreover, the spectra of the latter kind appear as more realistic to us because they can reproduce observations without resorting to scale-dependent biasing. Our APM-like simulations are in excellent agreement with earlier results, confirming the validity of the Juszkiewicz et al. [34] ansatz for $v_{12}(r)$ and the conjecture that the appearance of the shoulder in the correlation function near the $\xi = 1$ transition is a feature of gravitational dynamics rather than a peculiarity of a particular set of initial conditions.

Using these results, we proposed two tests of the hypothesis that galaxies trace mass. The first of the tests is based on an obvious idea, that if $\xi(r) = \xi_g(r)$, the galaxy correlation function near $\xi_g = 1$ should exhibit properties similar to those of the matter correlation function. We examined the behavior of the correlation function, derived from the APM catalogue and found exactly the same features we knew earlier from N-body simulations, in particular the agreement between the two characteristic scales, $r_* \approx r_o$. It is difficult to imagine how such an agreement could happen by a mere coincidence, which would have to be the case if ξ_g is unrelated to ξ. The agreement between the two characteristic scales can be used to constrain the linear biasing factor for the APM catalogue to be within 20% of unity. This constraint agrees with an earlier limit, obtained from measurements of the three-point correlation function from the APM survey [19], [20].

The second test confronts the v_{12}, predicted by assuming that the APM galaxies are unbiased tracers of mass with direct measurements of v_{12}. The results are again consistent with $b \approx 1$ and a low density parameter, $\Omega_m \approx 0.3$, in agreement with the limits obtained from the velocity data alone [44].

We are impressed how well the observations are reproduced by the simple calculations based on the assumption that galaxies follow the mass distribution, at least on large (weakly nonlinear) scales. We are unable to constrain biasing models with a large number of free parameters, but their predictive power is questionable and one may ask: are such models falsifiable and therefore worth constraining?

Our results are by no means final, they are also less rigorous than one could wish because we are limited by the accuracy of the present observational data. New generation of catalogues promise a dramatic improvement on this front in the near future (for an excellent collection of reports on the state of the art in this field, see Colombi *et al.* [45]).

ACKNOWLEDGMENTS

One of the co-authors (RJ) who was fortunate to attend this excellent meeting would like to thank Jim Fry for making it happen. We also thank Carlton Baugh for providing us with his APM-like simulations as well as his estimate of $\xi_g(r)$, based on the APM survey. RJ thanks Ruth Durrer for important discussions regarding the inflection point in $\xi(r)$ and for her hospitality at the University of Geneva. This work was supported by a collaborative grant between the Polish Academy of Science and the Spanish Consejo Superior de Investigaciones Cientificas. We also acknowledge support by grants from the Polish Government (KBN grant No. 2.P03D.01719), the Swiss Tomalla Foundation, and from IEEC/CSIC and DGES(MEC) (Spain), project PB96-0925.

REFERENCES

1. STRAUSS, M. & J. WILLICK. 1995. Phys. Rep. **26**: 271.
2. HAMILTON, A.J.S. 1998. *In* The Evolving Universe, Hamilton D., Ed. Kluwer. Dordrecht. p. 185.
3. DAVIS, M., E. EFSTATHIOU, C.S. FRENK & C.D.M. WHITE. 1985. Astrophys. J. **292**: 371.
4. VITTORIO, N., R. JUSZKIEWICZ & M. DAVIS. 1986. Nature **323**: 132.
5. MADDOX, S.J., G. EFSTATHIOU, W.J. SUTHERLAND & J. LOVEDAY. 1990. Mon. Not. R. Astr. Soc. **242**: 43P.
6. GAZTAÑAGA, E. 1995. Astrophys. J. **454**: 561.
7. JENKINS, A. *et al.* (The Virgo Consortium). 1998. Astrophys. J. **499**: 20.
8. REES, M.J. 1999. Preprint: astro-ph/9912373.
9. PEEBLES, P.J.E. 1999. *In* Clustering at High Redshift, A. Mazure & O. Le Fevre, Eds. (astro-ph/9910234).
10. DEKEL, A. & O. LAHAV. 1999. Astrophys. J. **520**: 24.
11. SELJAK, U. 2000. Preprint: astro-ph/0004086.
12. SCOCCIMARRO, R., R. SHETH, L. HUI & B. JAIN. 2000. Astrophys. J. In preparation.
13. BLANTON, M., R. CEN, J.P. OSTRIKER & M.A. STRAUSS. 2000. Astrophys. J. **531**: 1.
14. SOMMERVILLE, R.S. *et al.* 2001. Mon. Not. R. Astr. Soc. **320**: 289.
15. JUSZKIEWICZ, R., F. BOUCHET & S. COLOMBI. 1993. Astrophys. J. **412**: L9.
16. BERNARDEAU, F. 1994. Astron. Astrophys. **291**: 697.
17. BERNARDEAU, F. 1994. Astrophys. J. **433**: 1.
18. GAZTAÑAGA, E. 1994. Mon. Not. R. Astr. Soc. **268**: 913.
19. GAZTAÑAGA, E. & J.A. FRIEMAN. 1994. Astrophys. J. **437**: L13.
20. FRIEMAN, J.A. & E. GAZTAÑAGA. 1999. Astrophys. J. **521**: L83.
21. FELDMAN, H.A., J.A. FRIEMAN, J.N. FRY & R. SCOCCIMARRO. 2000. Preprint: astro-ph/0010205.
22. HAMILTON, A.J.S. & M. TEGMARK. 2000. Preprint: astro-ph/0008392.

23. JUSZKIEWICZ, R., P.G. FERREIRA, H.A. FELDMAN, A.H. JAFFE & M. DAVIS. 2000. Science **287:** 109.
24. VAN WAERBEKE *et al.* 2001. Preprint: astro-ph/0101511.
25. FISCHER, P., *et al.* (the SDSS Collaboration).1999. Preprint:astro-ph/9912119.
26. PEEBLES, P.J.E. 1993. Principles of Physical Cosmology. Princeton University Press. Princeton, NJ.
27. BINNEY, J. & M. MERRIFIELD. 1998. Galactic Astronomy. Princeton University Press. Princeton, NJ.
28. SOMERVILLE, R.S. *et al.* 2000. Mon. Not. R. Astr. Soc. **320:** 289.
29. PEEBLES, P.J.E. 1980. The Large--Scale Structure of the Universe. Princeton University Press. Princeton, NJ.
30. FRY, J. 1996. Astrophys. J. **461:** L65.
31. DAVIS, M. & P.J.E. PEEBLES. 1977. Ap. J. S. **34:** 425.
32. FISHER, K.B., M. DAVIS, M. STRAUSS, A. YAHIL & J. HUCHRA. 1994. Mon. Not. R. Astr. Soc. **267:** 927.
33. JUSZKIEWICZ, R., K. FISHER & I. SZAPUDI. 1998. Astrophys. J. **504:** L1.
34. JUSZKIEWICZ, R., V. SPRINGEL & R. DURRER. 1999. Astrophys. J. **518.** L25.
35. BAUGH, C.M. & E. GAZTAÑAGA. 1996. Mon. Not. R. Astr. Soc. **280:** 37.
36. GOTT, J.R. & M.J. REES. 1975. Astron. Astrophys. **45:** 365.
37. DAVIS, M. & P.J.E. PEEBLES. 1977. Astrophys. J. Suppl. **34:** 425.
38. VILLUMSEN, J. & M. DAVIS. 1986. Astrophys. J. **308:** 499.
39. LOKAS, E., R. JUSZKIEWICZ, F.R. BOUCHET & E. HIVON. 1996. Astrophys. J. **467:** 1.
40. SCOCCIMARRO, R. & J. FRIEMAN. 1996. Astrophys. J. **473:** 620.
41. GUZZO, L. 1997. New Astronomy **2:** 517.
42. BAUGH, C.M. 1996. Mon. Not. R. Astr. Soc. **282:** 1413.
43. WILLICK, J. *et al.* 1997. Ap. J. S. **109:** 333.
44. JUSZKIEWICZ, R., R. DURRER & P. FERREIRA. 2000. *In* Energy Densities in the Universe. Proc. Recontres de Moriond. In press .
45. COLOMBI, S., Y. MELLIER & B. RABAN, EDS., 1998. Wide Field Surveys in Cosmology. Editions Frontieres. Paris.

Peculiar Velocity Surveys: Optimal Moments Analysis

HUME A. FELDMAN,[a] RICHARD WATKINS,[b] ADRIAN L. MELOTT,[a] AND PATRICK GORMAN[a]

[a]*Department of Physics & Astronomy, University of Kansas, Lawrence, Kansas 66045*

[b]*Department of Physics, Willamette University, Salem, Oregon 97301*

ABSTRACT: A new formalism to analyze peculiar velocity surveys is presented. Results from these surveys are shown to be dominated by small-scale noise, aliasing, and incomplete cancellations. The formalism allows us to filter out the signal from scales that are not of interest and thus provides us with a clean signal that probe large scales. We use maximum likelihood techniques to analyze the filtered data and compare it to the analysis of the full dataset. The filtered analysis gives a much better parameter estimation than the full analysis, leading us to conclude that, indeed, the large-scale signal is obscured by small-scale noise.

KEYWORDS: Cosmology: distance scales, large-scale structure of the universe, observation, theory; Galaxies: kinematics and dynamics, statistics

1. INTRODUCTION

Although in principle, galaxy velocity field holds great promise as a direct probe of the underlying mass distribution, in practice the extraction of the information is difficult and fraught with both observational and theoretical pitfalls. Observationally, velocity surveys have irregular geometries and their boundary conditions are not usually well known. Further, they are (compared to redshift surveys) somewhat shallow and sample the volume discretely, nonuniformly, and sparsely. Theoretically, the mapping from velocity to density is complicated by nonlinear effects which necessitates various approximation schemes (e.g., [1]). Indeed, small-scale aliasing and incomplete cancellations [2]–[4] introduce spurious noise which masquerades as large-scale signal. These effects are difficult to disentangle and thus the resulting information is unreliable [3], [5], [6]. See FIGURE 1 for an example of the problems in analyzing velocity surveys.

In FIGURE 1a we plotted some theoretical power spectra and one from the IRAS–QDOT survey [7]. In FIGURE 1b we show the trace of the squared tensor window function of the bulk flow for various surveys. From this figure it is clear that, except on the largest scales, the different samples probe the power spectrum in very different ways; indeed, if one were to look at the full three-dimensional window function,

---

Address for correspondence: Hume A. Feldman, Department of Physics and Astronomy, University of Kansas, Lawrence, Kansas 66045. Voice: 785/864-4740; fax: 785/864-5262.

feldman@ku.edu

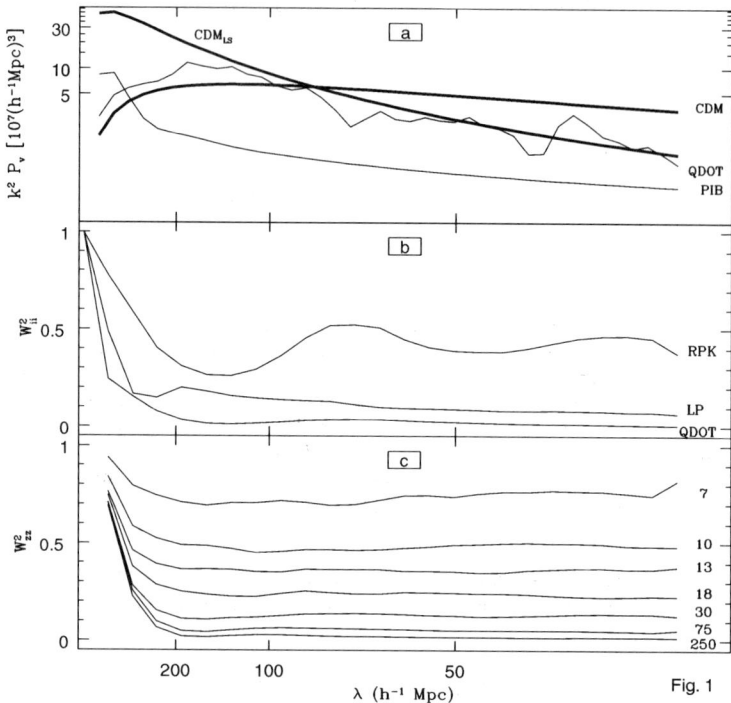

FIGURE 1. (a) The redshift corrected power spectra. (b) The trace of the squared tensor window functions for the bulk flow of the SNIa survey [8], Brightest Cluster Galaxy survey [9] and for comparison the IRAS–QDOT survey [7]. (c) The window functions for SNIa like surveys of different sizes. The contributions from large k fall as the number of data points increase.

the overlap between the two window functions would appear to be even smaller. This implies that while both vectors will have similar contributions from the very largest scales, contributions from smaller scales will in general not be correlated.

In FIGURE 1c we show the trace of the squared tensor window function of the bulk flow for mock surveys with the same distribution in redshift as a typical whole sky survey. The window functions are shown for samples of varying sizes, averaged over 20 realizations for each size. Averaging over many realizations tends to smooth out features associated with specific placements of sample objects, making the underlying "plateau" more prominent.

2. LIKELIHOOD METHODS FOR PECULIAR VELOCITIES

To analyze the observed line-of-sight velocities we assume that N objects with positions r_i and observed line-of-sight velocities v_i can be modeled as

$$v_i = \vec{v}(\vec{r}_i) \cdot \hat{r}_i + \delta_i \tag{1}$$

where $v(\vec{r}_i)$ is the linear velocity field and δ_i is the noise which also accounts for the deviations from linear theory. Assume the noise is Gaussian with variance $\sigma_i^2 + \sigma_*^2$ where σ_i is the observational error and σ_* is the contribution from nonlinearity and other things we neglected (see [5] for detailed analysis). The covariance matrix can be written as

$$R_{ij} = \langle v_i v_j \rangle = R_{ij}^{(v)} + \delta_{ij}(\sigma_i^2 + \sigma_*^2) \tag{2}$$

where

$$R_{ij}^{(v)} = \langle \vec{v}(\vec{r}_i) \cdot \hat{r}_i \; \vec{v}(\vec{r}_j) \cdot \hat{r}_j \rangle \tag{3}$$

In linear theory we can express the velocity power spectrum in terms of the density power spectrum and thus rewrite Eq. (3) as

$$R_{ij}^{(v)} = \frac{H^2 f^2(\Omega_0)}{2\pi^2} \int P(k) W_{ij}^2(k) dk . \tag{4}$$

The covariance matrix is a convolution of the density power spectrum and the squared tensor window function:

$$W_{ij}^2(k) = \frac{1}{4\pi} \int d\theta \, d\phi \, \sin\theta \, W_i(k) W_j^*(k) . \tag{5}$$

where

$$W_i(k) = (A^{-1})_{ij} \sum_n \frac{\hat{k} \cdot \hat{r}_n \, F_{n,j} \, e^{i\vec{k} \cdot \vec{r}_n}}{\sigma_n^2 + \sigma_*^2} . \tag{6}$$

and

$$A_{ij} = \sum_n \frac{F_{n,i} F_{n,j}}{\sigma_n^2 + \sigma_*^2} . \tag{7}$$

$F_{n,i}$ is the ith component of the unit vector of the nth galaxy.

The probability distribution for the line-of-sight peculiar velocities is

$$L(v_1, ..., v_N; P(k)) = \sqrt{|R^{-1}|} \exp\left(\frac{-v_i R_{ij}^{-1} v_j}{2}\right) . \tag{8}$$

Alternately, given a set of velocities $(v_1, ..., v_N)$ we can have $L(v_1, ..., v_N; P(k))$ to denote the likelihood functional for the power spectrum. Given a power spectrum parameterized by some vector $\Theta = (\theta_1, ..., \theta_s)$ then $L(v_1, ..., v_N; \Theta)$ is the likelihood functional for the parameter Θ. The value of the parameter vector that maximizes the likelihood is Θ_{ML}.

Given a set of true parameters Θ_0, we want a maximum likelihood estimator $\langle \Theta_{ML} \rangle = \Theta_o$. Then Θ_{ML} will vary over different realizations of $(v_1, ..., v_N)$. We may characterize our parameters with the mean $(\langle(\theta_{ML})_i\rangle)$ and the variance $[\Delta(\theta_{ML})^2{}_i = \langle(\theta_{ML})^2{}_i\rangle - \langle(\theta_{ML})_i\rangle^2]$. In the limit of large N: $(\langle \Theta_{ML} \rangle = (\Theta_o)_i)$ and variances are minimal.

Define the Fisher transformation matrix:

$$F_{ij} = \left\langle \frac{\partial^2(-\ln L)}{\partial \theta_i \partial \theta_j} \right\rangle \bigg|_{\Theta = \Theta_0}. \tag{9}$$

Variances for unbiased estimators are:

$$\Delta(\theta_{ML})_i \geq (F_{ii})^{-1/2}, \tag{10}$$

which is the so-called Cramér–Rao inequality. In the limit of large N this becomes an equality, here we assume that this limit is satisfied. If the velocities are Gaussianly distributed then the maximum likelihood estimator Θ_{ML} is unbiased. However, actual peculiar velocities contain non-Gaussian contributions. Nonlinear contributions will lead to Θ_{ML} being biased in an unpredictable way. In order to recover an unbiased estimator we utilize data compression methods. We use these methods to filter out unwanted information.

3. PROCEDURE

We start with a set of line-of-sight velocities v_i and positions r_i. We then calculate the covariance matrix R_{ij} and the window function $W_{ii}{}^2$. We must choose a power spectrum parameterized by $\Theta = (\theta_1, ..., \theta_s)$. Once we have these we can construct the likelihood functional $L(v_1, ..., v_N; \Theta)$ and then calculate the Fisher transformation matrix F_{ij}. Before we undertake this procedure we must discuss the data compression formalism we utilize (see references [16] and [17]).

4. DATA COMPRESSION

Replace N original line-of-sight velocities $v_1, ..., v_N$ with M moments $u_1, ..., u_M$ where $M \leq N$. Here we concentrate on linear data compression where the moments are:

$$u_i = B_{ij} v_j, \tag{11}$$

where B_{ij} is an $M \times N$ matrix. If $M < N$ then we lose information. However, we arrange it such that the lost information is primarily associated with small scales. Consider a CDM (cold dark matter) type power spectrum in which the nonlinear power is proportional to a single parameter θ_q. Given $v_1,\ldots,v_N$ determine the value of θ_q within a minimum variance $\Delta\theta_q^2 = 1/F_{qq}$ (the qq element of the Fisher matrix). The variance is the measure of the sensitivity of the data set to nonlinear scales. The larger the variance the less small-scale information the data set contains. Suppose we compress all the velocity information into a single moment.

$$u = b_i v_i, \qquad (12)$$

where b_i is a $1 \times N$ set of coefficients. The Fisher matrix for compressed data is:

$$\tilde{F}_{qq} = \frac{1}{2}\left|b_i \frac{\partial R_{ij}}{\partial \theta_q} b_j\right|^2, \qquad (13)$$

where our normalization scheme is $b_i R_{ij} b_j = 1$. Since $\Delta\theta_q^2 = 1/F_{qq}$ we can find a moment that carries the minimum information about θ_q by minimizing the right-hand side of Eq. (13). Introduce a Lagrange multiplier and extremize with respect to b_i to get

$$\left(\frac{\partial R_{ij}}{\partial \theta_q}\right) b_j = \lambda R_{ij} b_j . \qquad (14)$$

Since R_{ij} is symmetric and positive definite we can Cholesky decompose it:

$$R_{ij} = L_{ik} L_{jk} \qquad (15)$$

for some invertible matrix L_{ij}. Now we have an eigenvalue problem:

$$\left(L_{ki}^{-1} \frac{\partial R_{ij}}{\partial \theta_q} L_{lj}^{-1}\right)(L_{ml} b_m) = \lambda (L_{jk} b_j) \qquad (16)$$

Solving this gives us a set of N orthogonal eigenvectors $L_{ji}(b_n)_j$ with corresponding eigenvalues λ_n and moments $u_n = (b_n)_i v_i$. One can show that

$$\frac{1}{\Delta\theta_q} = \frac{|\lambda|}{\sqrt{2}}. \qquad (17)$$

Finding λ_n gives us the error bar of θ_q.

The moments u_n are statistically independent and of unit variance:

$$\langle u_n u_m \rangle = \delta_{ij}. \qquad (18)$$

We assume that v_i and thus u_n are Gaussian variables then the u_n are statistically independent. Thus, if we convert the velocities into N moments there will be no loss of information and the transformation matrix will be invertible. Also, since the moments are statistically independent, when we compress the data by removing selected moments, the information contained by those moments will be completely removed from the data.

5. MOMENT SELECTION

Arrange the moments in order of increasing eigenvalues:

$$\lambda_1 \leq \lambda_2 \leq \ldots \leq \lambda_N. \tag{19}$$

Each moment carries successively more information about θ_q with u_N and carries the maximum possible amount of information. Thus, our goal is to produce a data set that is less sensitive to the value of θ_q and keep as many moments as possible to retain the information about large scales. The question is how should we choose M? We use the following procedure: find the error $\Delta\theta_q$ that we can put on θ_q using the compressed data. The Fisher information matrix for the compressed data reduces to:

$$\tilde{F}_{qq} = \frac{1}{2} \sum_{n=1}^{M} \lambda_n^2. \tag{20}$$

The error bar is thus:

$$\Delta\theta_q = \sqrt{2} \left(\sum_{n=1}^{M} \lambda_n^2 \right)^{-1}. \tag{21}$$

Hence, we add the sum of the squares of the smallest eigenvalues until the desired sensitivity is reached. We use the following criterion:
1. Estimate $\theta_q = \theta_{q0}$.
2. Keep the largest number M such that $\Delta\theta_q \geq \theta_{q0}$.

If our estimate of θ_q is correct, then the set of moments $u_1, \ldots, u_M$ will not contain enough information to distinguish θ_q from zero.

6. POWER SPECTRUM MODEL

Assume that:

$$P(k) = P_1(k) + \theta_q P_{nl}(k), \tag{22}$$

where P_1 and P_{nl} are the linear and nonlinear power spectra, respectively. We assume that

$$P_1(k) = 0 \text{ for } k > k_{nl}, \tag{23}$$

$$P_{nl}(k) = 0 \text{ for } k < k_{nl}. \tag{24}$$

Use the BBKS (Bardeen, Bond, Kaiser, and Szalay [10]) power spectrum for the linear power spectrum $P_1(k)$. We can approximate the nonlinear power spectrum $P_{nl}(k) = P_0$ for $k_{nl} < k < k_c$ for some critical k. We also tried other parameterizations, for example, $P_{nl} \propto k^{-1}$ [11]. Where contribution of nonlinear scales to line-of-sight velocity dispersion (σ_*) should equal the estimate from the data. The "true" value for θ_q is $\theta_{q0} = 1$. We can express P_0 in terms of σ_* because

$$\sigma_* = 4\pi \int_{k_{nl}}^{k_c} P_v(k) k^2 dk = \text{constants} \int_{k_{nl}}^{k_c} P_{nl}(k) dk \qquad (25)$$

For a constant P_{nl} we get

$$P_0 = \frac{\sigma_*}{\text{constants}(k_{nl} - k_c)}$$

7. CREATING REALISTIC CATALOGS

The simulations used here for testing are done by N-body, specifically PM (particle mesh) method. This method is quite fast and, with a mean particle density of one per simulation cell, represents the maximum resolution that can be achieved without introducing two-body scattering that decouples the result from its initial conditions on small scales [12], [13]. The simulations used here all had 256^3 particles, and could be completed on an HP C240 workstation in about 24 hours apiece.

Initial conditions were generated by FFT with random phase perturbations. As behavior at high density peaks is not of interest here, the simulations were begun with an rms density fluctuation of 0.25 at the resolution limit.

Matter-dominated Friedman–Robertson–Walker background cosmologies were assumed, with the cosmological constant set to zero. Since nonlinear modes are filtered out, and the dynamical effect of nonzero Ω_λ is well understood in perturbation theory [14], we did not use nonzero values. The simulations are scaled by size and Hubble constant. The value of Ω_m, the matter density, determines the coupling of the scale factor to the particle dynamics in the usual way.

In these runs, the box size is taken to be 512 Mpc and the Hubble constant $h = H/100$ km s^{-1} Mpc^{-1} = 0.75. Thus the box size is redshift space corresponds to a diameter of 38,400 km s^{-1}. Formally, in the simulation h only sets an overall time scale; since both the expansion rate and particle velocities scale with h, the redshift space appearance does not change with h, only its overall scale.

The efficacy of the method in various backgrounds is tested by varying Ω in the range 0.25 to 1. One set had $\Omega_0 = 1$; the other set had output taken at various moments corresponding to a variety of values of $\Omega \geq 0.25$ and σ_8 (the mass density fluctuation after smoothing with a tophat window of radius 800 km s^{-1}) of $0.5 < \sigma_8 < 1$.

The parameter Γ determines the shape of the initial power spectrum. Smaller values of Γ are favored today, because they push the turnover in the slope of the power spectrum to large scales in better agreement with data. Again, to test the method we use values of $\Gamma = 0.25$, 0.5, and 1. Normally Γ is taken to be $\Omega_0 h$, as the turnover scale is set by the horizon at the end of the radiation dominated era. However, we break this assumed coupling and take Γ as a free parameter merely descriptive of spectral shape to provide a more complete test of the method. Both Γ and Ω vary independently.

We have three values of Γ; three moments per simulation at which data was taken corresponding to varying σ_8; and a low or high Ω_0 background, making a total of 18 configurations to analyze.

From the simulation there are now galaxies with positions and velocities. Other properties of the galaxies must be assigned to each in order to simulate a survey. The parameters which are important for this simulation are D (the diameter in km s^{-1} arcmin), B (the absolute blue magnitude), I (the absolute I-band magnitude), and the line-width η ($\eta = \log w - 2.5$). The procedure used to assign these parameters is based on the monté carlo simulations and selection effects by Freudling et al. (see reference [15] for the SFI catalog). If the redshift (z) is between 500 km s^{-1} and 10000 km s^{-1} the four properties are assigned to the galaxy. The parameter to assign first is the diameter. All other parameters can be calculated based on the diameter.

The diameters are distributed according to the following probability distribution:

$$P(D) = \frac{1}{D^*} e^{\frac{D_{min} - D}{D^*}} \quad (27)$$

where $D^* = 2231$ km s^{-1} arcmin, $D_{min} = 1500$ km s^{-1} arcmin, and $D_{max} = 30000$ km s^{-1} arcmin. Once the diameter is assigned to a galaxy, the most probable value for B can be found from the relation:

$$B(D) = a \log D + b \quad (28)$$

where $a = -4.87$ and $b = -0.05$. η can then be assigned using the following $\eta - B$ relationship:

$$\eta(B) = \frac{B - b_{tf} - b_c}{a_c + a_{tf}} - 2.5 \quad (29)$$

with $a_{tf} = -7.48$, $b_{tf} = -2.53$, $a_c = 1.78$, $b_c = -2.54$. Finally, the I-Band magnitudes are assigned from the following equation:

$$I = a_{tf} * \log w + b_{tf} \quad (30)$$

where $\log w = \eta + 2.5$. The errors for the B, η, and I (dB, $d\eta$, and dI, respectively) are calculated using known procedures.

A selection now must be made to create catalogs of approximately 1000 galaxies each. The selections are made based on a set of criterion to make the catalogs as realistic as possible. We then choose an η limits, a blue magnitude limit, and some angular and redshift masks. The parameters for these depend on the type of survey one wants to simulate. Finally, a selection is made based on the angular diameter versus the redshift of each galaxy. The angular diameter (A in arcmin) is calculated and chosen appropriately. From the galaxies which satisfy the above criterion, approximately 1000 are randomly selected to create the simulated catalogs.

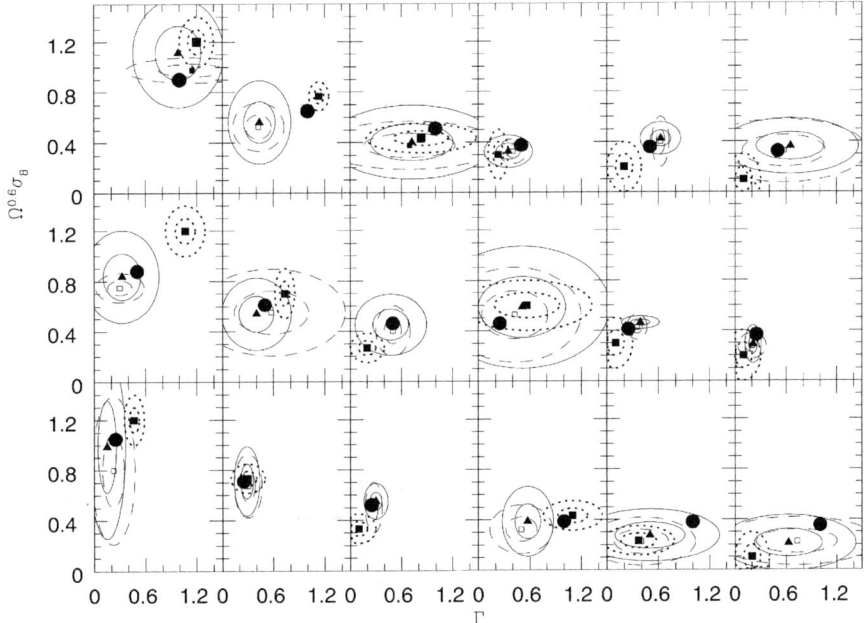

FIGURE 2. Comparison of parameter estimation for the mode analysis (*solid with solid triangle*) full analysis (*short dash with solid square*) and mode analysis including realistic distance errors (*long dash with open square*). The *large dot* is the true value of the parameters.

The Malmquist bias correction must now be applied to the simulated survey to correct the peculiar velocities of the galaxies from the simulation. This is done by correcting the distances. The corrected distance (d_{cor}) is calculated using the following equation.

$$d_{\text{cor}} = \frac{\int_{500}^{10000} r^3 n(r) e^{\frac{-(\ln(r/d_{\text{raw}}))^2}{2\Delta^2}}}{\int_{500}^{10000} r^2 n(r) e^{\frac{-(\ln(r/d_{\text{raw}}))^2}{2\Delta^2}}} \tag{31}$$

In Eq. (31), d_{raw} is the distance to the galaxy from the original simulation, $n(r)$ is the radial distribution of the galaxies approximated with a chi-squared fit, and Δ is the fractional error in the *I*-band magnitude (dI/I). The error in the distance (δ) is then calculated by taking the difference between the corrected distance and the distance from the original simulation

$$\delta = d_{\text{cor}} - d_{\text{raw}} \tag{32}$$

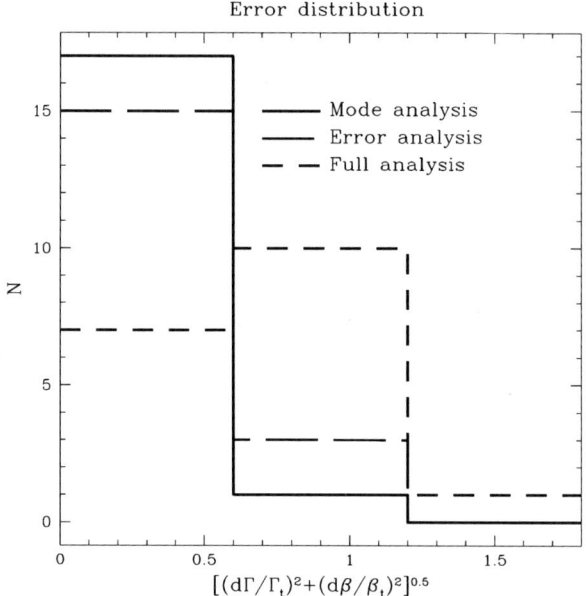

FIGURE 3. The distribution of absolute errors for the three possibilities as in FIG. 2.

The bias (b) is then calculated for a galaxy by taking the average of δ of all galaxies within a radius of 500 km s^{-1} from the galaxy. The b is then subtracted to get the final estimate of the distance to the galaxy (d). The peculiar velocity (v_{pec}) can then be found from the difference between the redshift and the final estimate of the distance.

$$v_{\text{pec}} = z - d \tag{33}$$

8. RESULTS

We constructed eight SFI-like surveys from the each of the eighteen parameterized simulations. Each survey has ≈1000 galaxies with redshift and angular positions and line-of-sight peculiar velocities. We filtered out the nonlinear modes. We found the maximum likelihood parameters for the power spectrum parameterized by [$\beta = \Omega^{0.6}\sigma_8$, Γ] and their variances. We then compared results to the true values known from the simulations. We did the same thing for the full analysis, that is, no filtering of the small-scale modes. In FIGURE 2 we see the results of the analysis. The error ellipses are the 1- and 2-σ errors derived only from the sample variance. To compare absolute errors in different analyses, in FIGURE 3 we have a histogram of the number of simulations (out of eighteen) for which the formalism gave us results $\sqrt{\beta^2 + \Gamma^2} <$

0.6. Clearly, the mode analyses, with or without errors, give much better results than the full analysis.

ACKNOWLEDGMENT

Research supported in part by the NSF grant AST-0070702 at Kansas.

REFERENCES

1. ZEL'DOVICH. 1970. Astron. & Astrophys. **5:** 84.
2. BRIGHAM, E.O. 1974. The Fast Fourier Transform. Prentice Hall. Englewood Cliffs, NJ.
3. WATKINS, R. & H.A. FELDMAN. 1995. Astrophys. J. **453:** L73–76.
4. FELDMAN, H.A. & R. WATKINS. 1998. Astrophys. J. **494:** L129–132.
5. FELDMAN, H.A. & R. WATKINS. 1994. Astrophys. J. **430** L17–20.
6. STRAUSS, M., R. CEN, J.P. OSTRIKER, M. POSTMAN & T. LAUER. 1995. Astrophys. J. **444:** 507.
7. FELDMAN, H.A., N. KAISER J.A. PEACOCK. 1994. Astrophys. J. **426:** 23–37.
8. RIESS, A., W. PRESS & R. KIRSHNER. 1995. Astrophys. J. **438:** L17–20 ([8]).
9. LAUER, T. & M. POSTMAN. 1994. Astrophys. J. **425:** 418–438.
10. BARDEEN, J. M., J.R. BOND, N. KAISER & A.S. SZALAY. 1986. Astrophys. J. **304:** 15.
11. KLYPIN, A.A. & A.L. MELOTT. 1992. Astrophys. J. **399:** 397.
12. KUHLMAN, B., A.L. MELOTT & S.F. SHANDARIN. 1996. Astrophys. J. **470:** 641.
13. SPLINTER, R.J., A.L. MELOTT, S.F. SHANDARIN & Y. SUTO. 1998. Astrophys. J. **497:** 38.
14. LAHAV, O., M.J. REES, P.B. LILJE, & J. PRIMACK. 1991. Mon. Not. R. Astr. Soc. **251:** 128.
15. FREUDLING et al. 1995. ASTRON. J. **110:** 2.
16. KENDALL, M.G. & A. STUART. 1969. The Advanced Theory of Statistics, Vol. 2. Grifin. London.
17. TEGMARK, M., A.N. TAYLOR & A.F. HEAVENS. 1997. Astrophys. J. **480:** 22–35.

Non-gaussianity Versus Nonlinearity of Cosmological Perturbations

LICIA VERDE

IfA, University of Edinburgh, Royal Observatory, Blackford Hill, EH9 3HJ, Edinburgh, U.K.

ABSTRACT: Following the discovery of the cosmic microwave background, the hot big-bang model has become the standard cosmological model. In this theory, small primordial fluctuations are subsequently amplified by gravity to form the large-scale structure seen today. Different theories for unified models of particle physics, lead to different predictions for the statistical properties of the primordial fluctuations, that can be divided in two classes: gaussian and non-gaussian. Convincing evidence against or for gaussian initial conditions would rule out many scenarios and point us toward a physical theory for the origin of structures.

The statistical distribution of cosmological perturbations, as we observe them, can deviate from the gaussian distribution in several different ways. Even if perturbations start off gaussian, nonlinear gravitational evolution can introduce non-gaussian features. Additionally, our knowledge of the Universe comes principally from the study of luminous material such as galaxies, but galaxies might not be faithful tracers of the underlying mass distribution. The relationship between fluctuations in the mass and in the galaxies distribution (*bias*), is often assumed to be local, but could well be nonlinear. Moreover, galaxy catalogues use the redshift as third spatial coordinate: the resulting redshift-space map of the galaxy distribution is nonlinearly distorted by peculiar velocities. Nonlinear gravitational evolution, biasing, and redshift-space distortion introduce non-gaussianity, even in an initially gaussian fluctuation field.

I investigate the statistical tools that allow us, in principle, to disentangle the above different effects, and the observational datasets we require to do so in practice.

KEYWORDS: Cosmology: large-scale structure, cosmic microwave background, galaxies, clusters

1. INTRODUCTION

Until recently in cosmology, non-gaussianity has been synonymous with nonlinearity; but, in the last 5 years or so, more and more objects like the galaxy of [1] at redshift 5.6 have been found. For the first time a galaxy has been found at higher redshift than the most distant known quasar. More recently, a galaxy at redshift almost 7 has been found [2]. The standard "inflationary" cosmological model with gaussian initial conditions predicts that these objects should be very rare. It is becoming in-

---

Address for correspondence: Licia Verde, Department of Astrophysical Sciences, Peyton Hall, Princeton University, Princeton, NJ 08544. Voice: +44-131-6688393; fax: +44-131-6688416.
lv@roe.ac.uk

creasingly difficult to accommodate the existence of so many high-redshift galaxies under the assumption that non-gaussianity is equivalent to nonlinearity, that is, postulating gaussian initial conditions. Non-gaussianity does not necessarily imply nonlinearity: there might be some primordial non-gaussianity and it is necessary to "find a way" to distinguish the two effects.

2. NON-GAUSSIANITY DUE TO NONLINEARITIES

Let us start by assuming gaussian initial conditions and investigate the effects of nonlinearities. We define the fractional density contrast δ as $\delta\rho/\rho$, where ρ is the mean density. The probability distribution of δ starts off symmetric around zero, with negligible tails for $|\delta| > 1$. Nonlinear gravitational evolution skews the distribution toward high densities: this is due to the fact that underdense regions cannot become more empty than the void ($\delta \geq -1$) while overdense regions can accrete matter arbitrarily (no upper limit on δ). This is not the only process that can skew an initially gaussian distribution. The mass in the Universe is mainly dark matter and cannot be observed directly—only galaxies can be observed, but mass and galaxy distributions may not be identical: the idea that galaxies are biased tracers of the mass distribution was introduced in the early eighties[1] [3] and has featured strongly in large scale structure (LSS) studies. In general, bias must alter the statistics of any underlying matter distribution, otherwise $\delta < -1$ for the galaxy field, which corresponds to a negative galaxy density. In different bias schemes suggested in the literature, the relation between the galaxy and the mass fluctuation fields (δ_g and δ, respectively) has been taken to be local, nonlocal, eulerian, lagrangian, stochastic, etc.

In what follows we will assume that $\delta_g(\mathbf{x}) = F[\delta(\mathbf{x})]$, that is the bias, is a local eulerian function of the underlying mass field. Furthermore we will assume (following [5]) that F can be expanded in Taylor series and we will truncate the expansion to the quadratic term:

$$\delta_g(\mathbf{x}) = b_0 + b_1\delta(\mathbf{x}) + \frac{b_2}{2}\delta^2(\mathbf{x}) + \dots \qquad (1)$$

b_0 is unimportant and simply ensures that $\langle\delta_g\rangle = 0$. This nonlinear operation on the matter field introduces some skewness, i.e., some non-gaussianity. As first suggested by Fry [6] it is possible to disentangle the two non-gaussian contributions, nonlinear gravity and bias, by looking at higher-order correlations in the mildly nonlinear regime. In particular, if the initial fluctuations are gaussian and cosmological structures grow by gravitational instability, the three-point correlation function is intrinsically a second-order quantity[2] and is detectable in the mildly nonlinear regime. If then bias can be expressed as in Eq. (1), it is possible to show that a likelihood analysis[3] of the bispectrum (the three-point correlation function in Fourier space) can yield b_1 and b_2.

---

[1]Although the fact that galaxies of different morphologies have different spatial distributions and they cannot all be good tracers of the underlying mass distribution, was known much before the introduction of the concept of bias (e.g., [4]).
[2]In the quantity δ, assumed to be small.

The bispectrum $B(\mathbf{k}_1,\mathbf{k}_2,\mathbf{k}_3)$ is defined as

$$\langle \delta_{\mathbf{k}_1}\delta_{\mathbf{k}_2}\delta_{\mathbf{k}_3}\rangle = (2\pi)^3 B(\mathbf{k}_1, \mathbf{k}_2, \mathbf{k}_3)\delta^D(\mathbf{k}_1 + \mathbf{k}_2 + \mathbf{k}_3) \quad (3)$$

where δ_k is the Fourier transform of $\delta(\mathbf{x})$. Due to the presence of the Dirac delta function δ^D, the bispectrum can be non zero only when the three $\mathbf{k}$ form a triangle.

In practice, the higher-order statistic (the bispectrum) exploits the fact that gravitational instability skews the density field as it evolves, creating sheets and filament-like structures reminiscent of the Zeldovich pancakes. Nonlinear bias also introduces skewness but does so by shifting the iso-density contours up and down, without modifying the shape of the structures. These two effects can be disentangled by using different triangle shapes for the bispectrum.

There are several advantages in performing this sort of analysis in Fourier space, most of them are the same advantages of the power spectrum over the two-point correlation function. We will recall here only the following: the estimates of power on different scales can be made uncorrelated; it is easy to deal with the error estimate; and more importantly it is easy to distinguish between linear, mildly nonlinear and highly nonlinear scales. In the mildly nonlinear regime the bispectrum is given by:

$$\langle \delta_{g,\mathbf{k}_1}\delta_{g,\mathbf{k}_2}\delta_{g,\mathbf{k}_3}\rangle = (2\pi)^3 P_g(k_1)P_g(k_2)[c_1 J(\mathbf{k}_1, \mathbf{k}_2) + c_2]\delta^D(\mathbf{k}_1 + \mathbf{k}_2 + \mathbf{k}_3) \\ + cyc \quad (3)$$

where P_g denotes the galaxy power spectrum, $J(\mathbf{k}_1,\mathbf{k}_2)$ is a known function of the two k-vectors and

$$c_1 = \frac{1}{b_1} \quad c_2 = \frac{b_2}{b_1^2}. \quad (4)$$

In the absence of bias ($c_1 = 1$, $c_2 = 0$) it would be easy to isolate the non-gaussianity generated by gravitational instability. Nevertheless, even in the presence of bias, it is possible to disentangle gravity from bias. Equation (3) is in a form suitable for a likelihood analysis for the two bias parameters via c_1 and c_2 [9], once the covariance is known.[4] A generating functional approach to calculate *analytically* the N-point function and therefore the covariance for the bispectrum was introduced in [10], [11]. The performance of the method has been tested on biased and unbiased N-body simulations,[5] with very promising results[6] [11] for the application to forthcoming galaxy redshift surveys such as SDSS and 2dF. An estimation for the expected error achievable from present galaxy surveys, yields an error on c_1 of about 100% [11] and is therefore not particularly useful.

---

[3]The likelihood method can easily be generalized to measure the lagrangian [7] and stochastic (e.g., [8]) bias parameters.

[4]The bispectrum is a three point quantity, its covariance is a six-point quantity (pentaspectrum).

[5]The N-body simulation was provided by the Hydra-consortium and produced using the code [12].

[6]For a different approach see the work of J. Frieman.

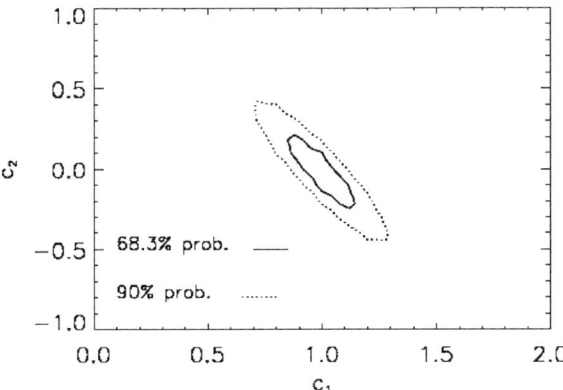

FIGURE 1. Likelihood for $c_1 = 1/b_2$ and $c_2 = b_2/b_1^2$, (where b_1 and b_2 are linear and quadratic bias parameters) for an unbiased simulation ($b_1 = 1$, $b_2 = 0$) in redshift space. The solid and dotted lines show the 1 and 3 σ confidence levels respectively. This figure shows that it is possible to disentangle the nonlinear gravitational instability, bias and redshift space distortion signals, and also that from future galaxy redshift surveys such as SDSS and 2dF, the bias could be known with an accuracy better that 10%.

3. REAL WORLD ISSUES

Of course, reality is always more complicated: in a realistic galaxy survey several complications arise due to the presence of shot noise and selection function, but more importantly due to redshift space distortions. Galaxy surveys in fact use the redshift as the third spatial coordinate. The redshift would be an accurate distance indicator in a perfectly homogeneous universe; but the Universe is clumpy, inhomogeneities perturb the Hubble flow and introduce peculiar velocities. The resulting redshift-space map of the galaxy distribution is thus distorted along the line of sight, and the nature of this distortion is intrinsically nonlinear. On large scales the coherent inflow into overdense regions introduces a squashing effect in the redshift map (the *great wall* effect), on smaller scales, the virialized highly nonlinear structures appear elongated along the line of sight (the *fingers-of-God*), heavily contaminating the mildly nonlinear regime were most of the signal for the bispectrum comes from. In [13] we showed that, with an accurate modeling of redshift space distortions in the distant observer approximation, it is possible to disentangle the effects of redshift-space anisotropies from the bias and gravitational effects. This is achieved by combining a second-order perturbation theory description of the coherent inflow (e.g,. [14]) with an exponential velocity dispersion model, and discarding the k-modes where the contamination from highly nonlinear structures is too big to be successfully modeled. An alternative approach has also been explored (e.g., [15]).

The result of the likelihood analysis performed on an redshift space unbiased simulation of 100 h^{-1} Mpc side is shown in FIGURE 1.

It shows not only that it is possible to disentangle the nonlinear gravitational instability, bias and redshift space distortion signals, but also that from future galaxy redshift surveys such as Sloan digital sky survey (SDSS) and the Anglo-Australian two-degree field (2dF), the bias could be known with an accuracy better that 10%. Before achieving this goal however, there are several other issues to deal with, that are potentially serious for any Fourier based technique. In particular, these are the mask, the difficulty of obtaining redshifts for close pairs of galaxies, the holes arising from bright-star drills, and the variable completeness. We have investigated and quantified these effects on a simulated catalogue and concluded that a 10% error on the bias parameter could be achieved from the 2dF survey [16].

3.1. Bypassing the Redshift-Space Distortions

Two-dimensional surveys avoid the redshift-space distortion problems, but, in principle, contain less information. However, because of the smaller observational effort required, these can contain a much larger number of objects. For example the APM survey at present contains 10^6 galaxies, the DPOSS catalogue will have 50 million galaxies and the SDSS will provide us with a two-dimensional map of 10^7 galaxies.

To treat the projection of higher-order correlations on the celestial sphere, the spherical nature of the distribution cannot be ignored. Spherical harmonics are eigenfunctions for the two-dimensional surface of the sphere and therefore are the natural basis describing a two-dimensional random field on the sky. When comparing with Fourier space analysis we have that

$$\delta_{\mathbf{k}} \longrightarrow a_\ell^m \tag{5}$$

and in particular for the bispectrum

$$\delta^D(\mathbf{k}_1 + \mathbf{k}_2 + \mathbf{k}_3) \longrightarrow \begin{pmatrix} \ell_1 & \ell_2 & \ell_3 \\ m_1 & m_2 & m_3 \end{pmatrix} \tag{6}$$

where on the RHS we have the three-J symbol. To perform the same sort of analysis as the one illustrated in sections 2 and 3 an expression that relates the 3D bispectrum to the projected one in spherical harmonics is needed:

$$\langle a_{\ell_1}^{m_1} a_{\ell_2}^{m_2} a_{\ell_3}^{m_3} \rangle = \begin{pmatrix} \ell_1 & \ell_2 & \ell_3 \\ m_1 & m_2 & m_3 \end{pmatrix} \left[\frac{1}{\bar{n}} \frac{16}{\pi} \sqrt{\frac{(2\ell_1+1)(2\ell_2+1)(2\ell_3+1)}{4\pi^3}} \times \right.$$

$$\int dk_1 dk_2 i^{\ell_1+\ell_2} k_1^2 k_2^2 \Psi_{\ell_1}(k_1) \Psi_{\ell_2}(k_2) \sum_{\ell \ell_6 \ell_7} i^{\ell_6+\ell_7} (-1)^\ell B_\ell(k_1, k_2)(2\ell_6+1)(2\ell_7+1)\bar{\rho}$$

$$\left. \times \int dr r^2 \psi(r) j_{\ell_6}(k_1 r) j_{\ell_7}(k_2 r) \begin{pmatrix} \ell_1 & \ell_6 & \ell \\ 0 & 0 & 0 \end{pmatrix} \begin{pmatrix} \ell_2 & \ell_7 & \ell \\ 0 & 0 & 0 \end{pmatrix} \begin{pmatrix} \ell_3 & \ell_6 & \ell_7 \\ 0 & 0 & 0 \end{pmatrix} \begin{Bmatrix} \ell_1 & \ell_2 & \ell_3 \\ \ell_7 & \ell_6 & \ell \end{Bmatrix} \right. \tag{7}$$

$$\left. + cyc \right]$$

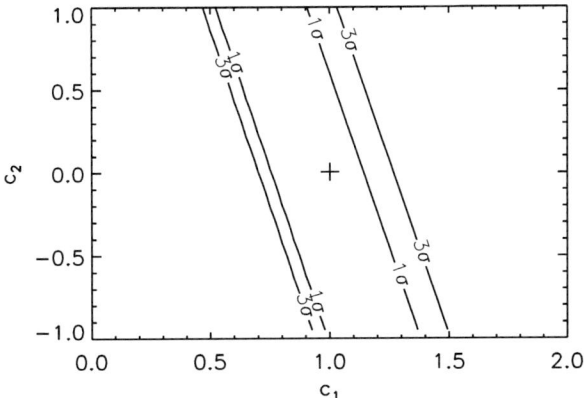

FIGURE 2. Likelihood contours for degenerate triplets configurations. The two levels are the 1–σ and 3–σ confidence levels and the + indicates where the true value for the parameters lies. Perturbation theory breaks down at $\ell \sim 50$. Adding other configurations does not help much: this is about the best result obtainable from projected surveys.

This expression is quite complicated: the derivation and the detailed explanation of it can be found in [17]. For the purpose of this contribution we only have to notice that it is an *exact* expression relating the spherical harmonics projected bispectrum

$$\langle a_{\ell_1}^{m_1} a_{\ell_2}^{m_2} a_{\ell_3}^{m_3} \rangle$$

to the 3D bispectrum expressed through its Legendre coefficients $B_\ell(\mathbf{k}_i \mathbf{k}_j)$. $\Psi_{\ell_i}(k_j)$ is a known function of the selection function, j_ℓ denotes the spherical Bessel function and $\{\ldots\}$ denotes the Wigner 6-J symbol.

By performing a likelihood analysis to measure the bias parameter on a all sky simulation with the APM selection function, we find that the results are not encouraging for projected catalogues (see FIGURE 2).

It is therefore preferable to undertake the bispectrum study of 3D galaxy redshift surveys such as SDSS and 2dF using the method described above. The good news is that the exact expression for the projected bispectrum in spherical harmonics has applications in a variety of areas such as cosmic microwave background (CMB) and gravitational lensing studies.

4. PRIMORDIAL NON-GAUSSIANITY: CMB VERSUS LSS

Up to now we have assumed gaussian initial conditions. However, among the theories for structure formation, only inflation predicts a nearly gaussian distribution for the primordial fluctuations, with deviations from gaussianity which are calculable, small, and dependent on the specific inflationary model (e.g., [18], [19], [20]). In other models such as nonstandard inflation or topological defects models initial conditions are non-gaussian. CMB and LSS data will shortly improve dramatically:

it is therefore timely to ask which of the CMB or LSS will provide a better probe of the nature of primordial fluctuations. The advantage of looking at CMB maps is that the fluctuation distribution should be close to the primeval form, but the disadvantage is that the amplitude of fluctuations is small and there are foregrounds and other effects to account for. The advantages of looking at LSS is that the signal has been amplified by gravity. This is, however, also a disadvantage because gravity skews the distribution. Nonlinear gravity, bias, and redshift space distortions might completely swamp the primordial signal. Following [21], as a discriminating statistic we will use the bispectrum, but we will start by considering the skewness as an example to illustrate some of the effects. The skewness is defined as

$$S_3 = \frac{\langle \delta^3 \rangle}{\langle \delta^2 \rangle^2} \tag{8}$$

For a gaussian field the skewness is zero, while for an initially gaussian field evolved under gravitational instability to second order in δ, the skewness becomes 34/7 and is constant in time.[7] In what follows for CMB related calculations, we will assume an Einstein de Sitter Universe, this assumption is justified because we shall be concerned with factors of 10 while the cosmology can change the results only by factors of order unity. Suppose the initial conditions are very close to gaussian, but with a small primordial skewness, parameterized by $S_3^P(z = 1100)$ at recombination. The effect on LSS will be (e.g. [22]):

$$S_3(z = 0) = S_{3,0}^P + 34/7 + \mathcal{F} \tag{9}$$

where $\mathcal{F}$ includes a complicated dependence on the three- and four-point function that we will ignore for the moment and $S_{3,0}$ scales as:

$$S_{3,0}^P = \frac{S_3^P(z)}{(1+z)}, \tag{10}$$

the primordial skewness redshifts away. We can then make a thought experiment: assume we know the real space position of every particle in the whole Hubble volume. The smallest error for the skewness, that is the smallest $S_{3,0}^P$ detectable, on 20 h^{-1} Mpc scales is $S_{3,0}^P \sim 10^{-2}$, which implies $S_3^P(z = 1100) \sim 10$. We can repeat the exercise for the CMB where now, for consistency, we consider the smallest detectable skewness on 0.2° scales, obtaining $S_3^P(z = 1100) \sim$ few < 10.

This example already shows that CMB seems to be more sensitive to primordial deviation from gaussianity than LSS, but now we will proceed more accurately by considering the bispectrum — in fact the bispectrum contains more information than the skewness and has all the advantages of being a Fourier space quantity. In the absence of bias, the LSS bispectrum in second-order perturbation theory for nongaussian initial condition is:

---

[7]The value 34/7 is strictly true only for an Einstein de Sitter Universe, but the skewness does not depend strongly on cosmological parameters.

$$B(\mathbf{k}_1,\mathbf{k}_2,\mathbf{k}_3) = B_0(\mathbf{k}_1,\mathbf{k}_2,\mathbf{k}_3) + 2J(\mathbf{k}_1,\mathbf{k}_2)P(k_1)P(k_2) + cyc$$
$$+ \int d^3k J(\mathbf{k}', \mathbf{k}_3 - \mathbf{k}')T^c(\mathbf{k}', \mathbf{k}_3 - \mathbf{k}', \mathbf{k}_1, \mathbf{k}_2) + cyc. \quad (11)$$

B_0 is the primordial bispectrum evolved linearly and corresponds to $S_{3,0}$ of Eq. (9); the second term is the usual gravitational instability bispectrum and corresponds to the 34/7 term in Eq. (9); T^c denotes the Fourier counterpart of the connected four point function and the integral term corresponds to $\mathcal{F}$ of Eq. (9). We then parameterize the LSS bispectrum as:

$$B = P(k_1)P(k_2)[2J(\mathbf{k}_1, \mathbf{k}_2)c_1 + c_2] + cyc \quad (12)$$

because we know how to estimate c_1 and c_2 from LSS studies (Section 2). In the very idealized case where the real space position of every particle in the SDSS volume was known, the minimum c_1 and c_2 detectable would be, respectively, $c_1 \sim 10^{-3}$ and $c_2 \sim 10^{-2}$ ignoring all the real world complications of shot noise, selection function, etc.

On the CMB side, the bispectrum for realistic non-gaussian models is given by (e.g., [23]):

$$B_{\ell_1 \ell_2 \ell_3} = f(\ell_1, \ell_2, \ell_3)\alpha [C_{\ell_1} C_{\ell_2} + cyc] \quad (13)$$

where α is the amplitude, f is a known function of ℓ and C_ℓ denotes the CMB power spectrum. The minimum error σ_α on the amplitude α can be obtained via the Fisher information matrix:

$$\sigma_\alpha^{-2} = -\left\langle \frac{\partial^2 \ln \mathcal{L}}{\partial \alpha^2} \right\rangle \cong \sum_{\ell_1 \leq \ell_2 \leq \ell_3} \frac{(B_{\ell_1 \ell_2 \ell_3}|\alpha=1)^2}{n C_{\ell_1} C_{\ell_2} C_{\ell_3}} \sum_{m_1 m_2 m_3} \frac{\begin{pmatrix} \ell_1 \ell_2 \ell_3 \\ m_1 m_2 m_3 \end{pmatrix}}{N(m_i, \ell_i)} \quad (14)$$

where $n = 1/2$ and $N(m_i, \ell_i)$ is the number of nonzero terms like $C_{\ell_1} C_{\ell_2} C_{\ell_3}$ in the covariance and ranges from 1 to 30. Here we neglect partial sky coverage effects, and by constraining $\ell \lesssim 100$ pixel noise and small angular scale effects are negligible.

In [21] we investigated the LSS and CMB bispectrum as a discriminating statistic for several physically motivated non-gaussian models. There is an infinitude of deviations from gaussianity and one cannot address them all: we thus restrict ourselves to physically motivated models where the non-gaussianity can be dialed from zero (the gaussian limit) and is assumed to be small. In particular we consider the non-gaussianity parameterized by:

$$\Phi = \phi + \varepsilon(\phi^2 - \langle \phi^2 \rangle) \quad (15)$$

where ϕ denotes a gaussian field and for the moment we will assume that Φ is the gravitational potential; the non-gaussianiy parameter is ε, that is zero for a gaussian field, ~1 for standard inflation, but can be as big as ~20 for some nonstandard inflationary models. The CMB effect is given by $2\varepsilon/A_{SW} = \alpha$ where A_{SW} is the Sachs–Wolfe coefficient ~1/3.

It is possible to see that [21], if the CMB distribution from the future satellite missions turns out to be consistent with gaussian, the smallest ε allowed would be ~20. The LSS effect is $c_2 = b_2/b_1^2 + 10^{-6}\,\varepsilon$ and the T^c contribution is negligible. By substituting the minimum ε measurable from CMB in this expression and ignoring bias, we obtain $c_2 = 10^{-4}$ — about two orders of magnitude smaller that the minimum c_2 detectable from LSS even in the most idealized conditions.

In [21] also, other physically motivated non-gaussian models have been considered, the result is qualitatively always the same: *if future CMB maps are consistent with the gaussian hypothesis then any non-gaussianity seen in the LSS bispectrum is due to nonlinear gravity or bias*, and we know how to disentangle the two.

5. LOOKING AT SMALLER SCALES

In practice, CMB studies can be affected by noise and foreground and, more importantly, there might be models in which non-gaussianity is present mainly on LSS or galaxy scales, which are not fully accessible with CMB experiments. I will therefore investigate another two ways to detect primordial non-gaussianity on scales smaller than CMB ones.

5.1. Detecting Non-Gaussian Initial Conditions from Large-scale Structure

It is possible to bypass the contamination due to nonlinear clustering and discriminate between gaussian and non-gaussian initial conditions by using higher-order statistics in LSS studies such as the trispectrum — that is, the connected four-point correlation function in Fourier space [24]. This quantity has the advantage of having a rather simple growth rate, with no complicating contributions from nonlinear gravity in second-order perturbation theory and that the analysis depends on cosmology and bias only through the measurable quantity[8] β. The departures from gaussian statistics can be parameterized by introducing the quantity H, which is the fractional excess of the 4-point function over the gaussian (disconnected) trispectrum. This quantity, in specific cases, can give us a meaningful measure of "non-gaussianity": for mildly non-gaussian fields, a *measurement* of H can reliably be made. For highly non-gaussian fields, the gaussian hypothesis can be rejected, but the measurement of H will be unreliable. Following [24] it is possible to deal with redshift-space distortions, biasing, spatially varying selection function and shot-noise. FIGURE 3 shows the minimum χ^2 analysis for the parameter H from a redshift-space unbiased CDM-like simulation. By applying this method to future galaxy surveys, such as the SDSS, it will be possible to place tight constraints on initial departures from gaussian behavior.

---

[8]The quantity β arises naturally when studying the large-scale squashing effect of structures in redshift-space maps. It is defined as $\beta \simeq \Omega_0^{0.6}/b$ where b is the linear bias parameter.

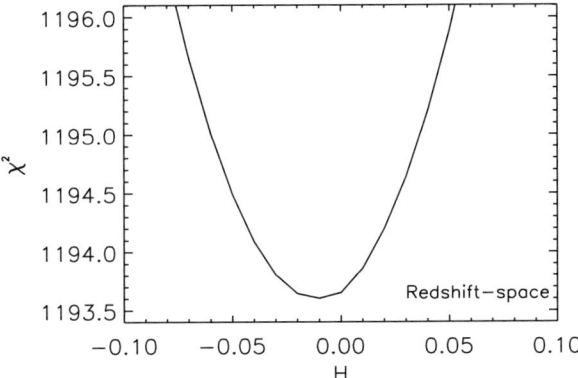

FIGURE 3. Minimum χ^2 analysis for the parameter H from a redshift-space unbiased CDM-like simulation. The analysis is largely bias independent. The true value for H is 0, and it is nicely within the 1–σ level (minimum $\chi^2 + 0.5$.). The quantity H, in specific cases, can give us a meaningful measure of "non-gaussianity": for mildly non-gaussian fields.

5.2. The Abundance of High-redshift Objects as a Probe of Non-gaussian Initial Conditions

Large-scale structure probes scale much larger than galaxies but smaller than those accessible by CMB observations, and probes the present-day Universe at $z = 0$; conversely, CMB maps probe the Universe at redshift $z \sim 1100$ and even larger scales. The abundance of cosmological structures at redshift in between these two ends and in particular at $z > 1$, also contains vital information about the nature of primordial fluctuations due to the fact that one is probing the tail of the distribution. FIGURE 4 shows that the effect of a small non-gaussianity is dramatic on the tails: high peaks are amplified much more than low ones and deep through can become local maxima.

To extract this information, the Press–Schechter (PS) formalism [25] needs to be extended to non-gaussian initial conditions. The PS is an analytical model to calculate the *mass function* (that is, the number of object per unit mass at a given redshift per unit volume), within an appropriate theoretical model (see R. Sheth and S. Shandarin contributions in this volume). The key ingredient is the probability density function (PDF) for the smoothed dark matter field $\mathcal{P}(\delta_M)$; in fact, the number density of objects above a given mass M (corresponding to a smoothing radius R) at a given redshift (the *mass function*) is proportional to the quantity $P_{>\delta_c}$:

$$P_{>\delta_c}(\delta_M) = \int_{\delta_c}^{\infty} \mathcal{P}(\delta_M) d\delta_M$$

where δ_c is the threshold overdensity for the object to collapse, $P_{>\delta_c}(\delta_M)$ is evidently a function of the redshift of formation (or collapse) of the object z_c, this redshift de-

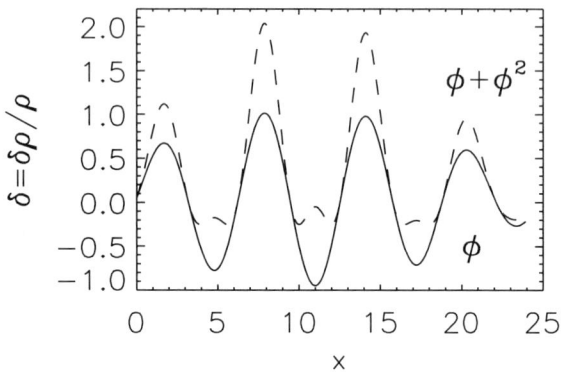

FIGURE 4. The effect of a small non-gaussianity are dramatic on the tail of the distribution. Assume φ here denotes a one-dimensional gaussian fluctuation filed (*black line*); the *dashed line* is given by φ + φ². High peaks are amplified much more than low ones and deep through can become local maxima.

pendence is enclosed in δ_c: $\delta_c(z) = \Delta_c/D(z)$. Here $D(z)$ is the linear growth factor, which in turn depends on the background cosmology, and Δ_c is the linear extrapolation of the overdensity for spherical collapse; Δ_c is traditionally taken to be ~1.68 but other values have also been used (see S. Shandarin and R. Sheth contributions, this volume).

For gaussian fields, $\mathcal{P}(\delta_M)$ is of course well known, but needs to be computed for non-gaussian initial condition.

Given a physically motivated parameterization of primordial non gaussianity, we set off to calculate the PDF for the *smoothed* dark matter field analytically. Of course one could evaluate the PDF from numerical simulations, but this approach is plagued by the difficulty of properly accounting for the nonlinear way in which resolution and finite box-size effects propagate into the statistical properties of the non-gaussian field.[9]

As before we parameterize the non-gaussianity as in Eq. (15) where Φ can be the potential or the density field. To properly deal with the smoothing we use a path-integral approach in the calculation of the PDF:

$$\mathcal{P}(\delta_R) = \left\langle \delta^D\left[\phi_R(x) + \varepsilon \int d^3y F_R(|x-y|)\phi^2(y) - C - \delta_R(x)\right] \right\rangle \quad (17)$$

---

[9]For example imagine computing the power spectrum of a non-gaussian field $\psi = \phi + \phi^2 - \langle\phi^2\rangle$ where φ is gaussian with a power-law power spectrum P_ϕ. The power spectrum of ψ involves computing the convolution of two P_ϕ. This convolution is an integral over k from 0 to infinity. When performing the operation on a simulation box the final result would be as if the integral was truncated at $k \geqslant 2\pi/L$ and $k \geqslant 2\pi/l$ where L is the side of the box and l is the grid resolution.

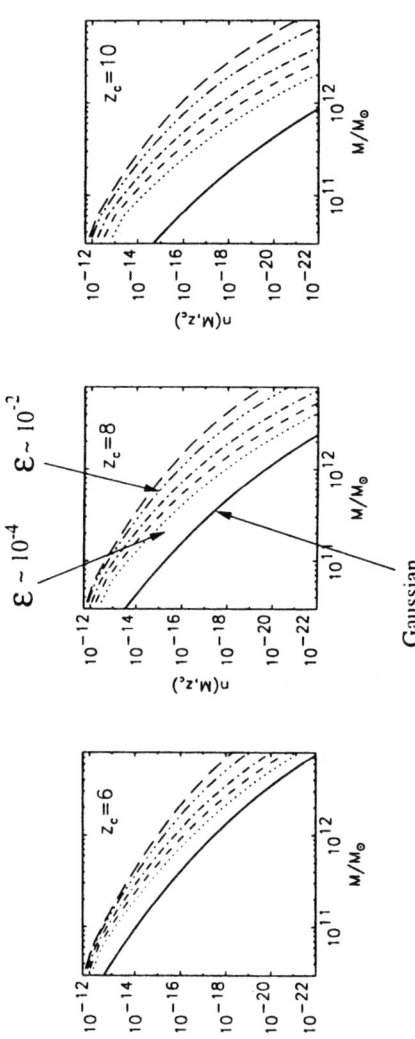

FIGURE 5. Effects of the non-gaussianity of Eq. (15) applied on the density fluctuation field. The figure shows only galaxy mass scales: the solid line is the mass function for gaussian initial conditions, all the other lines have non-gaussianity parameter $10^{-4} \leq \varepsilon \leq 10^{-2}$. For $z_c = 8$, a small non-gaussianity parameter $\varepsilon \sim 10^{-4}$ increases the mass function by two orders of magnitude at about $M \sim 10^{10} M_\odot$.

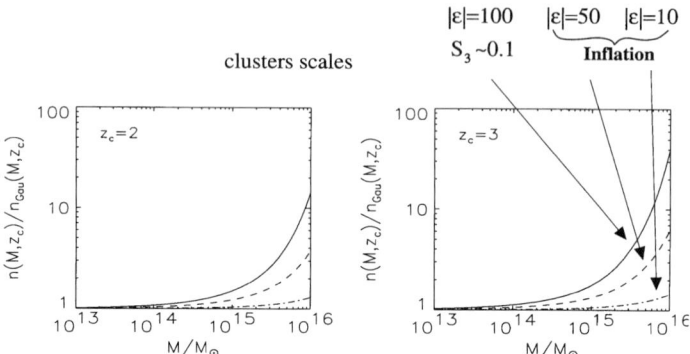

FIGURE 6. The non-gaussianity of Eq. (15) applied to the gravitational potential field has dramatic effects on cluster scales. In particular observations of clusters with $z_c \sim 2\text{--}3$ and $M > 10^{15}\,M_\odot$ could put constraints on inflationary models. The absolute value for ε here is relatively big, but the deviation from gaussianity is still small: the skewness here is of the same order of magnitude as in FIGURE 5.

where the R subscript denotes the smoothed quantity, $\delta^D[\ldots]$ the Dirac delta function, $\langle \ldots \rangle$ the ensemble average, and F is defined through its Fourier transform ($\tilde{F}_R(k) = W_R(k)T(k)g(k)$ with W_R the smoothing, T the transfer function and $g = 1$ for the density or $g = -2/3(k/H_0)^2 \Omega_{0,m}^{-1}$ for the potential). In ϕ_R, smoothing and transfer functions are easily accounted for, but in the non-gaussian part, the presence of the smoothing in $F_R(|x - y|)$ makes the quantity nonlocal. The Dirac delta function can be expressed in its integral representation and the ensemble average can be written as an integral over all ϕ configurations weighted by the gaussian probability density functional. In this way we can express an unknown quantity in terms of all known quantities and we are able to obtain the non-gaussian PDF for the smoothed field analytically. The details of the derivation can be found in [26]. The main result is that, for mildly non-gaussian initial conditions with small positive skewness $S_{3,R}$, the threshold for collapse δ_c is lowered, in particular [26]:

$$\delta_c(z_c) \longrightarrow \delta_c(z_c)\left[1 - \frac{S_{3,R}}{3}\delta_c(z_c)\right] \tag{18}$$

where $S_{3,R}$ is directly proportional to the non-gaussianity parameter ε of Eq. (15). By lowering the threshold for collapse rare objects will form more easily, and this has a huge impact on the tails of the distribution as shown in FIGURES 5 and 6.

In particular, FIGURE 5 (middle panel) shows that the non-gaussianity of Eq. (15) applied to the density field with $\varepsilon \sim 10^{-4}$, can change the number density of objects of mass $\sim 10^{11} M_\odot$ that collapse at redshift $z_c = 8$, by two orders of magnitude. Conversely, it is clear from FIGURE 5.2 that non-gaussian mapping 15, applied to the potential field, has dramatic effects on cluster scales. Observations of clusters with $z_c \gtrsim 2$ and $M \gtrsim 10^{15} M_\odot$ could put some constraints on inflationary models [27].

5.2.1. A worked example

Up to date 6 galaxies with confirmed spectroscopic redshifts have been observed with redshifts $5 < z < 7$. The observed comoving density N is for a $\Omega_{0,m} = 0.3$, $\Lambda = 0.7$ (ΛCDM) Universe is $N \geq 8.3 \times 10^{-4} (h^{-1} \text{Mpc})^{-3}$, $[N \geq 3.6 \times 10^{-4} (h^{-1} \text{Mpc})^{-3}$ for an Einstein–de Sitter Unverse]. Their masses are very uncertain, but some estimate can be obtained with simple arguments about their observed star formation rate, these estimates can then be compared with Ly$_\alpha$ width observations [26]. The gaussian ΛCDM model predicts $N \geq 5.2 \times 10^{-5} (h^{-1} \text{Mpc})^{-3}$, a factor ~20 fewer objects, while the Einstein–de Sitter model predicts $N \geq 10^{-7} (h^{-1} \text{Mpc})^{-3}$ — a factor 10^4 fewer objects! Only $\varepsilon \sim 10^{-3}$ in the density is needed to reconcile ΛCDM predictions with observations. Alternatively, this discrepancy could be explained postulating an error of about a factor 4 in the mass determination. As larger telescopes such as NGST get on line it will be possible to determine masses more accurately and thus constrain the amount of primordial non-gaussianity on galaxy scales.

6. CONCLUSIONS

We have shown that non-gaussianity does not necessarily mean nonlinearity, but it is possible to distinguish different kinds of nonlinearity, for example, bias, gravitational evolution, redshift space distortions. For the physically motivated non-gaussian models we considered, it turns out that CMB bispectrum is better that LSS bispectrum to detect primordial non-gaussianity: if the future CMB missions will produce maps that are consistent with the gaussian hypothesis, any non-gaussianity seen in the LSS bispectrum can be unambiguously attributed to the effects of non-linearities. Thus, if this is the case, from on-going LSS surveys such as SDSS and 2dF we will be able to know the bias within a few percent of accuracy. We have also shown that, to measure the bias parameter, ongoing 3D surveys are much better than 2D ones, even with full sky coverage, but the method developed has applications in different areas such as CMB and gravitational lensing studies. To conclude, we have seen different ways to disentangle primordial non-gaussianity from effects of nonlinearity: CMB bispectrum, LSS trispectrum, and the abundance of high-redshift objects such as galaxies and clusters. These methods probe the Universe at different scales and at different times and in addition they are sensitive to different moments of the distribution. We should therefore conclude that these methods are complementary rather than mutually exclusive.

ACKNOWLEDGMENTS

I would like to thank my collaborators in this work Alan Heavens, Sabino Matarrese, Marc Kamionkowski, Limin Wang, and Raul Jimenez. I also would like to thank the organizers for a very enjoyable workshop.

REFERENCES

1. WEYMANN, R.J., D. STERN, A. BUNKER, H. SPINRAD, F.H. CHAFFE, R.I. THOMPSON & L.J. STORRIE-LOMBARDI. 1998. Keck Spectroscopy and NICMOS Photometry of a Redshift $Z = 5.60$ Galaxy. Astrophys. J. Lett. **505:** L95–L98.
2. CHEN H.W., K.M. LANZETTA & S. PASCARELLE. 1999. Spectroscopic identification of a galaxy at a probable redshift of $Z = 6.68$. Nature **398:** 586–588.
3. KAISER, N. 1984. On the spatial correlations of Abell clusters. Astrophys. J. Lett. **284:** L9–L12.
4. HUBBLE, E.P. 1938. The Realm of Nebulae. Yale University Press. New Heaven.
5. FRY, N.J. & E. GAZTANAGA. 1993. Biasing and hierarchical statistics in large-scale structure. Astrophys. J. **413:** 447–452.
6. FRY, N.J. 1994. Gravity, bias, and the galaxy three-point correlation function. Phys. Rev. Lett. **73:** No. 2, 215–219.
7. CATELAN P., F. LUCCHIN, S. MATARRESE & C. PORCIANI. 1998. The bias field of dark matter haloes. Mon. Not. R. Astr. Soc. **297:** 692–712.
8. TARUYA A., K. KOYAMA & J. SODA. 1999. Quasi-nonlinear evolution of stochastic bias. Astrophys. J. **510:** 541–550.
9. VERDE, L. 1996. Large Scale Bias in the Universe. Laurea Thesis, University of Padua. Padua.
10. VERDE, L., A.F. HEAVENS & S. MATARRESE. 1997. Measuring Ω_0 via the bias parameter. *In* Generation of Cosmological Large-Scale Structure, D.N. Scramm & P. Galeotti, Eds: 245–250. Nato Asi series. Kluwer Academic. Dordrecht.
11. MATARRESE, S., L. VERDE & A.F. HEAVENS. 1997. Large-scale bias in the Universe: bispectrum method. Mon. Not. R. Astr. Soc. **290:** 651–662.
12. COUCHMAN, H.M.P., P.A. THOMAS & F.R. PEARCE. 1995. Hydra: an adaptive-mesh implementation of P 3M-SPH. Astrophys. J. **452:** 797–813.
13. VERDE L., A.F. HEAVENS, S. MATARRESE & L. MOSCARDINI. 1998. Large-scale bias in the Universe II: redshift space distortions. Mon. Not. R. Astr. Soc. **300:** 747–756.
14. HEAVENS, A.F., S. MATARRESE & L. VERDE. 1998. The nonlinear redshift-space power spectrum of galaxies. Mon. Not. R. Astr. Soc. **301:** 797-808.
15. SCOCCIMARRO, R., H.M.P. COUCHMAN & J.A. FRIEMAN. The bispectrum as a signature of gravitational instability in redshift space. Astrophys. J. **517:** 531–540.
16. VERDE, L. 2000. Ω_0, bias and primordial non-Gaussianity. Ph.D. thesis, University of Edinburgh. Edinburgh.
17. VERDE L., A.F. HEAVENS & S. MATARRESE. 2000. Projected bispectrum in spherical harmonics and its applications to angular galaxy catalogues. Mon. Not. R. Astr. Soc. Submitted, astro-ph/0002240.
18. FALK T., R. RANGARAJAN & M. SREDNICKI. 1993. The angular dependence of the three-point correlation function of the CMB radiation as predicted by inflationary cosmologies. Astrophys. J. Lett. **403:** L1–L5.
19. GANGUI A., F. LUCCHIN, S. MATARRESE & S. MOLLERACH. 1994. The three-point correlation function of the cosmic microwave background in inflationary models. Astrophys. J. **430:** 447-456.
20. WANG, L. & M. KAMIONKOWSKI. 2000. The cosmic microwave background bispectrum and inflation. Phys. Rev. D **61:** 063504.
21. VERDE L., L. WANG, A.F. HEAVENS & M. KAMIONKOWSKI. 2000. Large-scale structure, cosmic microwave background and primordial non-Gaussianity. Mon. Not. R. Astr. Soc. **313:** 141–147.
22. FRY, J.N. & R.J. SHERRER. 1994. Skewness in large-scale structure and non-Gaussian initial conditions. Astrophys. J. **429:** 36–42.
23. LUO, X. 1994. The angular bispectrum of the cosmic microwave background. Astrophys. J. Lett. **427:** L71–L74.
24. VERDE, L. & A.F. HEAVENS. 2000. On the trispectrum as a gaussian test for cosmology. Astrophys. J. In press.
25. PRESS W.H. & P. SCHECHTER. 1974. Formation of galaxies and clusters of galaxies by self-similar gravitational condensation. Astrophys. J. **187:** 425–438.

26. MATARRESE S., L. VERDE & R. JIMENEZ. 2000. The abundance of high-redshft objects as a probe of non-Gaussian initial conditions. Astrophys. J. **539:** astro-ph/0001366.
27. VERDE, L., R. JIMENEZ, M. KSMIONKOWSKI & S. MATARRESE. 2001. Tests for primordial non-gaussianity. Mon. Not. R. Astr. Soc. In press.

The Cosmological Mass Function in the Zel'dovich Approximation

SERGEI F. SHANDARIN

*Department of Physics and Astronomy, University of Kansas,
Lawrence, Kansas 66045, USA*

ABSTRACT: The Press–Schechter theory of the cosmological mass function and its modifications allow constraint of cosmological scenarios of structure formation. Recently, a few new models that explored the influence of anisotropic collapse on the shape of the mass function have been suggested. I discuss in more detail a particular model that assumes a fluid particle becomes part of a gravitationally bound halo when the smallest eigenvalue of the deformation tensor of the filtered initial density field reaches a certain threshold (like the filtered density contrast reaches the threshold in the Press–Schechter formalism). Choosing the smallest eigenvalue guarantees that the fluid particle in question experiences collapse along all three axes. The model shows a better agreement with the N-body simulations than with the standard Press–Schechter model.

KEYWORDS: Cosmological mass function; Galaxies; Gravitationally bound object; Clumps; Clusters

1. INTRODUCTION

The derivation of the distribution of masses in gravitationally bound objects is one of the principle goals of the theory of structure formation (for a review see [1]). Comparing the theoretical mass function with observations provides important constraints on cosmological models (see, e.g., [2]–[4]). Rich clusters of galaxies represent a particularly interesting class of objects for two reasons. First, they are the largest gravitationally bound objects in the universe and therefore represent rare events. As one of the consequences of being rare events clusters are particularly sensitive to some parameters of cosmological models (Ω_m and σ_8). Second, the formation of clusters is a relatively simple process since it is primarily determined by the gravitational dynamics, while other processes (hydro, thermal, etc.) are less important than, for example, in the process of galaxy formation. As a result, numerical simulations of clusters of galaxies are more realistic and more reliable than simulations of galaxy formation.

Measuring the mass function of galaxy clusters is not easy, but recently certain progress has been achieved for both optically (see, e.g., [5], [6]) and X-ray [7] selected samples. Although there are systematic differences between mass functions obtained by different groups there is a general agreement in a broad sense.

Address for correspondence: Sergei Shandarin, Department of Physics and Astronomy, University of Kansas, Lawrence, Kansas 66045, USA.Voice: 785/864-5274; fax: 785/864-5262.
sergei@ukans.edu

Most of theoretical derivations of the cosmological mass function are based on the ideas of Press and Schecter [8] and can be summarized as follows:

- The mass fraction ($F(>M)$) in gravitationally bound objects with masses greater than M can be estimated as the fraction of mass satisfying the collapse condition at this scale ($\Psi(\delta_M > \delta_c)$): $F(>M) = 2\Psi(\delta_M > \delta_c)$.
- The collapse condition is local, that is, it can be expressed in terms of the quantities at one point.
- The quantity that determines the collapse is the linearly extrapolated filtered density contrast $\delta_M \geq \delta_c$ at a given point.
- The value of the threshold $\delta_c = 3/20(12\pi)^{2/3} \approx 1.69$ corresponds to the collapse of a spherical top-hat model with similar initial density contrast. It was assumed that collapse of the spherical top-hat model approximately corresponded to virialization of the gravitationally bound clump.

The mathematical aspects of the Press–Schechter formalism are outlined in the following section. Here I discuss briefly some of ideas suggested since the formalism was proposed in 1974.

The excursion set approach [9], [10], justified the assumption that $\Phi(>M) = 2\Psi(\delta_M > \delta_c)$ in the case of a sharp k-space filter.

Many realized that the threshold $\delta_c = 1.69$ does not provide the best fit to N-body simulations. Although some authors used the canonical value (e.g., [10], [11]) others preferred the lower values: $\delta_c = 1.58$ [2], $\delta_c = 1.44$ [12], or even as low as $\delta_c = 1.33$ [13], [14]. Recently, Shapiro et al. [15] showed that virialization of the top-hat model occurs when linear extrapolation of density contrast reaches $\delta_c \approx 1.52$.

One of the major efforts in reduction of discrepancy of theory with simulations has been related to incorporating the anisotropic character of gravitational collapse. Bond and Meyers [2] developed a model that incorporated both the Zel'dovich approximation on large scales and the collapse of a homogeneous ellipsoid on the nonlinear scale. Monaco [16] suggested a different collapse condition that corresponded to collapse along the first axis in the Zel'dovich approximation. Audit et al. [17] incorporated some of the nonlinear effects into an anisotropic collapse model. Lee and Shandarin [18] suggested to use the collapse condition corresponding to collapse along all three axes as described by the extrapolation of the Zel'dovich approximation. Sheth and Tormen [19] obtained an analytic fit to the numerical mass function in the SCDM, OCDM, and ΛCDM models and then Sheth, Mo, and Tormen [20] provided a semianalytic derivation of the formula assuming an anisotropic collapse and incorporating some nonlocal effects. All but one of the models mentioned above assumed that the formation of a gravitationally bound object is related to collapse along three axes. Only Monaco [16] assumed the collapse condition corresponding to collapse along only the first axis.

In this paper I briefly review the Press–Schechter formalism and then describe the derivation of the mass function (λ_3-function) in the Zel'dovich approximation. I compare the result with the standard Press–Schechter model and the model suggested by Sheth, Mo, and Tormen [20]. I briefly discuss the results of comparison of the λ_3-function with N-body simulations. Finally, I discuss the effect of the initial gravitational potential on the cosmological mass function and show that clusters have a significant tendency to form in the troughs of the initial gravitational potential.

2. THE PRESS–SCHECHTER FORMALISM

The mass function $n(M)$ is the number density of gravitationally bound clumps with masses between M and $M + dM$. Let $F(>M)$ be the fraction of the mass contained in the gravitationally bound objects with masses greater than M. Press and Schechter [8] suggested the fraction $F(>M)$ and the mass function $n(M)$ can be related as

$$n(M) = -\frac{\bar{\rho}}{M}\frac{\partial F}{\partial M}, \qquad (1)$$

where $\bar{\rho}$ is the mean mass density in the universe and the minus sign reflects the fact that F is a decreasing function of M.

Press and Schechter also made the assumption that the fraction of mass $F(>M)$ can be estimated as a fraction of mass $\Psi(\delta_M > \delta_c)$ in the initial density field filtered with the window function W (corresponding to $\tilde{W}$ in k-space)

$$\delta_M(x, t) = D(t)\int \delta_{in}(x')W(|x' - x|/R) \sim d^3x', \qquad (2)$$

satisfying the collapse condition $\delta_M > \delta_c$. Here $\delta = (\rho - \bar{\rho})/\bar{\rho}$ is density contrast, $D(t)$ is linear growth factor, x is the comoving coordinate. The mass M and the linear scale R of the filter are related as

$$M = f_W \frac{4\pi}{3} R^3 \bar{\rho}, \qquad (3)$$

where f_W is a factor depending on the shape of the smoothing filter W. For a sharp k-space filter adopted here $f_W = 9\pi/2 \approx 14.1$ and thus $M = 6\pi^2 R^3 \bar{\rho}$ (see, e.g., [21]).

Assuming that the initial density contrast is a gaussian random field, its pdf (probability distribution function) is

$$p(\delta_M) = \frac{1}{\sqrt{2\pi}\sigma_M}\exp\left[-\frac{\delta_M^2}{2\sigma_M^2}\right], \qquad (4)$$

where the variance σ_M^2 is a function of mass M

$$\sigma_M^2 = \int \frac{d^3k}{(2\pi)^3} P(k)\tilde{W}^2(kR), \qquad (5)$$

where $P(k) = |\delta_k|^2$ is the initial spectrum of perturbations and $\tilde{W}(kR)$ is the window function in the k-space.

Press and Schechter argued that a fluid element becomes part of a gravitationally bound object of mass M when its linearly extrapolated density contrast δ_M reaches the critical value $\delta_c = 3/20(12\pi)^{2/3} \simeq 1.69$. This corresponds to collapse of the top-hat spherical perturbation, having initial density contrast similar to the fluid element in question. Collapse of a spherical top-hat model has been assumed to correspond approximately to virialization of the clump. Recently, Shapiro et al. [15] showed that virialization corresponds to $\delta_c \simeq 1.52$ rather than to $\delta_c \simeq 1.69$.

The fraction of mass satisfying the collapse condition on scale M is

$$\Psi(\delta_M > \delta_c) = \frac{1}{\sqrt{2\pi}\sigma(M)} \int_{\delta_c}^{\infty} \exp\left[-\frac{\delta_M^2}{2\sigma^2(M)}\right] d\delta_M$$

$$= \frac{1}{2}\mathrm{erfc}\left[\frac{\delta_c}{\sqrt{2}\sigma(M)}\right]$$

(6)

where erfc(x) is the complementary error function. Assuming that $F(M) \approx \Psi(\delta_M > \delta_c)$ one easily obtains the mass function $n(M)$ (Eq.1).

One obvious problem with this result is that the normalization integral

$$\int_0^{\infty} dF \approx \Psi(\delta_{M=\infty} > \delta_c) = \frac{1}{2},$$

(7)

meaning that only half the mass is contained in gravitationally bound clumps. Press and Schechter renormalized $n(M)$ by introducing an additional factor of 2 ($F(M) = 2\Psi(\delta_M > \delta_c)$)

$$n_{ps}(M) = -2\frac{\bar{\rho}}{M}\frac{\partial \Psi}{\partial M} = 2\frac{\bar{\rho}}{M}\frac{\partial \sigma}{\partial M}\frac{\partial \Psi}{\partial M}$$

$$= -\sqrt{\frac{2}{\pi}}\frac{\bar{\rho}}{M}\frac{d\sigma}{dM}\frac{\delta_c}{\sigma^2(M)}\exp\left[-\frac{\delta_c^2}{2\sigma_M^2}\right].$$

(8)

Later, the normalization problem was correctly resolved in the frame of the excursion set model [9], [10]. The derivation of Press and Schechter did not take into account the so-called cloud-in-cloud problem. Function $\Psi(\delta_M > \delta_c)$ in Eq. (6) gives the fraction of mass that satisfies the collapse condition at the filtering scale M. Some fluid particles, however, may satisfy the collapse condition at larger filtering scales. In the correct model, fluid elements must be assigned to clumps of mass M_1 being equal to the largest filtering mass at which the collapse condition is fulfilled. In the excursion set formalism this corresponds to the first crossing of the collapse threshold δ_c, while δ evolves with the growth of σ.

An elegant method to normalize the mass function was suggested by Jedamzik [22] (see also the discussion by Yano et al. [23]) who derived the integral equation Equation (9) relates the fraction of the fluid elements that first crossed the collapse threshold at the filtering scale $M_1(dM_1 n(M_1)M_1/\bar{\rho})$ and the fraction of mass satisfy-

$$\Psi(\delta_M > \delta_c) = \int_M^\infty dM_1 n(M_1) \frac{M_1}{\bar\rho} P(M, M_1). \tag{9}$$

ing the collapse condition at filtering scale $M(\Psi(\delta_M > \delta_c))$. Function $P(M, M_1)$ is the probability that a fluid particle first crossed the collapse threshold at the scale M_1 satisfies the collapse condition at the scale M. In the case of the sharp k-space filter and gaussian δ_M this probability is exactly equal to 1/2 for all $M_1 > M$. Thus, integral equation (9) can be immediately solved for the mass function $n(M)$. The solution is the correctly normalized mass function of Eq. (8).

3. MASS FUNCTION IN THE ZEL'DOVICH APPROXIMATION

The simplest theory describing the anisotropic character of the gravitational collapse in a generic case of random initial condition is the Zel'dovich approximation [24](see also [25] for a discussion). In particular, the Zel'dovich approximation provides a formula for an anisotropic collapse of a fluid element

$$\rho(\mathbf{q}, t) = \frac{\bar\rho}{[1 - D(t)\lambda_1(\mathbf{q})][1 - D(t)\lambda_2(\mathbf{q})][1 - D(t)\lambda_3(\mathbf{q})]}, \tag{10}$$

where $D(t)$ is the linear growth function and $\lambda_1(\mathbf{q})$, $\lambda_2(\mathbf{q})$, and $\lambda_3(\mathbf{q})$ are the eigenvalues of the initial deformation tensor. Using the ordering convention $\lambda_1(\mathbf{q}) > \lambda_2(\mathbf{q})$ and $\lambda_2(\mathbf{q}) > \lambda_3(\mathbf{q})$ the condition $1 - D(t)\lambda_1(\mathbf{q}) = 0$ has been interpreted as a collapse of a fluid particle along one principle axis [24]. Similarly, conditions $1 - D(t)\lambda_i(\mathbf{q}) = 0$ ($i = 2, 3$) can be interpreted as collapses along the second and third principle axes.

Shandarin and Klypin [26] showed that the formation of gravitationally bound clumps was best correlated with the maxima of the smallest eigenvalue (λ_3) of the initial deformation tensor. Although the formation of clumps also may be related to other pointlike singularities [27], here we assume that a fluid particle becomes a part of a gravitationally bound clump of mass M when its smallest eigenvalue λ_3 reaches the critical value λ_c at the largest filtering scale M [18]. The Zel'dovich approximation (Eq. 10) predicts that the collapse condition is $\lambda_c = 1$ (it is assumed that $D(t)$ normalized to $D(t_0) = 1$, where t_0 is the present time). However, because of multi-streaming effect all fluid particles (except the set of measure zero) enter the multi-stream flow regions before they collapse. We approximately incorporate this complex effect by reducing the threshold λ_c to a smaller value. The comparison with the Press–Schechter mass function as well as with the numerical mass function suggests that $\lambda_c = 0.37$ is a reasonable choice.

The derivation of the mass function in the Zel'dovich approximation is similar to the Press–Schechter derivation except that the collapse condition is $\lambda_3(M) = \lambda_c$ instead of $\delta_M = \delta_c$. Doroshkevich [28] derived the joint pdf of three eigenvalues

$$p(\lambda_1, \lambda_2, \lambda_3) = \frac{3375}{8\sqrt{5}\pi\sigma^6} \exp\left(-\frac{3I_1}{\sigma^2} + \frac{15I_2}{2\sigma^2}\right)(\lambda_1 - \lambda_2)(\lambda_2 - \lambda_3)(\lambda_1 - \lambda_3) \tag{11}$$

where $I_1 = \lambda_1 + \lambda_2 + \lambda_3$, $I_2 = \lambda_1\lambda_2 + \lambda_2\lambda_3 + \lambda_3\lambda_1$ and σ^2 is the density contrast variance as defined in Eq. (5). Integrating $p(\lambda_1, \lambda_2, \lambda_3)$ over two eigenvalues one can obtain the pdf of one of the eigenvalues. We are interested in the collapse along the third axis and therefore $p(\lambda_3)$ is of primary interest

$$p(\lambda_3) = \frac{\sqrt{5}}{12\pi\sigma}\left\{3\sqrt{3\pi}\exp\left(-\frac{15\lambda_3^2}{4\sigma^2}\right)\mathrm{erfc}\left(\frac{\sqrt{3}\lambda_3}{2\sigma}\right)\right.$$

$$+ \sqrt{2\pi}\left(20\frac{\lambda_3^2}{\sigma^2} - 1\right)\exp\left(-\frac{5\lambda_3^2}{2\sigma^2}\right)\mathrm{erfc}\left(\sqrt{2}\frac{\lambda_3}{\sigma}\right) \qquad (12)$$

$$\left. - 20\frac{\lambda_3}{\sigma}\exp\left(-\frac{9\lambda_3^2}{2\sigma^2}\right)\right\}$$

Repeating the derivation of the previous section using the pdf of Eq. (12) instead of Eq. (4) we have an analog of the normalization integral equation (Eq. 9):

$$\Psi(\lambda_3(M) > \lambda_c) = \int_M^\infty dM_1 n(M_1)\frac{M_1}{\bar{\rho}}P(M, M_1), \qquad (13)$$

where $\Psi(\lambda_3(M) > \lambda_c)$ is the fraction of mass where $\lambda_3(M) > \lambda_c$ on the filter scale M.

Solving exactly Eq. (13) is much more difficult than Eq. (9) because now $P(M,M_1)$ is not a constant. In the limit $M_1 - M \ll M$, the probability $P = 0.5$ as in Eq. (9), but in the limit $M_1 \gg M$ it drops to $P = 0.08$. Lee and shandarin [18] used the limiting value $P = 0.08$ and analytically derived the mass function in the Zel'dovich approximation

$$n(M) = \frac{25\sqrt{5}}{24\pi}\frac{\bar{\rho}}{MdM}\frac{d\sigma}{\sigma_M^2}\lambda_{3c}$$

$$\left\{3\sqrt{3\pi}\exp\left(-\frac{15\lambda_3^2}{4\sigma^2}\right)\mathrm{erfc}\left(\frac{\sqrt{3}\lambda_3}{2\sigma}\right)\right.$$

$$+ \sqrt{2\pi}\left(20\frac{\lambda_3^2}{\sigma^2} - 1\right)\exp\left(-\frac{5\lambda_3^2}{2\sigma^2}\right)\mathrm{erfc}\left(\sqrt{2}\frac{\lambda_3}{\sigma}\right) \qquad (14)$$

$$\left. - 20\frac{\lambda_3}{\sigma}\exp\left(-\frac{9\lambda_3^2}{2\sigma^2}\right)\right\}$$

The following sections compare the obtained result with the Press–Schechter and Sheth–Mo–Tormen mass functions as well as with numerical simulations.

4. COMPARISON OF THREE ANALYTIC MASS FUNCTIONS

Both the Press–Schechter and λ_3-mass functions have a common factor

$$\frac{\bar{\rho}\, d\sigma}{M\, dM}$$

which depends on the initial spectrum and $f(\sigma) \equiv \partial F/\partial \sigma$ that completely characterizes a model. Thus, comparing different models is convenient by comparing $f(\sigma)$ as functions of σ. The Press–Schechter and λ_3-functions are

$$f_{PS}(\sigma) = \sqrt{\frac{2}{\pi}} \frac{\delta_c}{\sigma^2} \exp\left(-\frac{\delta_c^2}{2\sigma^2}\right),$$

$$\begin{aligned} f_{\lambda_3}(\sigma) = \frac{25\sqrt{5}}{24\pi} \frac{\lambda_{3c}}{\sigma^2} \Bigg\{ & 3\sqrt{3\pi} \exp\left(-\frac{15\lambda_3^{\frac{2}{3}}}{4\sigma^2}\right) \mathrm{erfc}\left(\frac{\sqrt{3}\lambda_3}{2\sigma}\right) \\ & + \sqrt{2\pi}\left(20\frac{\lambda_3^{\frac{2}{3}}}{\sigma^2} - 1\right) \exp\left(-\frac{5\lambda_3^{\frac{2}{3}}}{2\sigma^2}\right) \mathrm{erfc}\left(\sqrt{2}\frac{\lambda_3}{\sigma}\right) \\ & - 20\frac{\lambda_3}{\sigma} \exp\left(-\frac{9\lambda_3^{\frac{2}{3}}}{2\sigma^2}\right)\Bigg\} \end{aligned} \quad (15)$$

Sheth et al. [19], [20] derived a new mass function that fits the results of the N-body simulations better:

$$f_{SMT}(\sigma) = A\left[1 + \left(\frac{a\delta_c^2}{\sigma^2}\right)^{-q}\right]\sqrt{\frac{2}{\pi}}\frac{\sqrt{a}\delta_c}{\sigma^2}\exp\left(-\frac{a\delta_c^2}{\sigma^2}\right). \quad (16)$$

Here $a = 0.707$, $q = 0.3$ and the constant $A = 0.322$ found from the normalization condition

$$\int_0^\infty f(\sigma) d\sigma = 1. \quad (17)$$

Choosing $a = 1$, $q = 0$, and the constant $A = 1$ one obtains the Press–Schechter function. These three functions are shown in FIGURE 1a. The small box shows the range of σ where the theoretical mass functions were checked against N-body simulations by Sheth and Tormen [19] and Lee and Shandarin [29]. FIGURE 1b shows the ratios f_{PS}/f_{SMT} and f_{λ_3}/f_{SMT}.

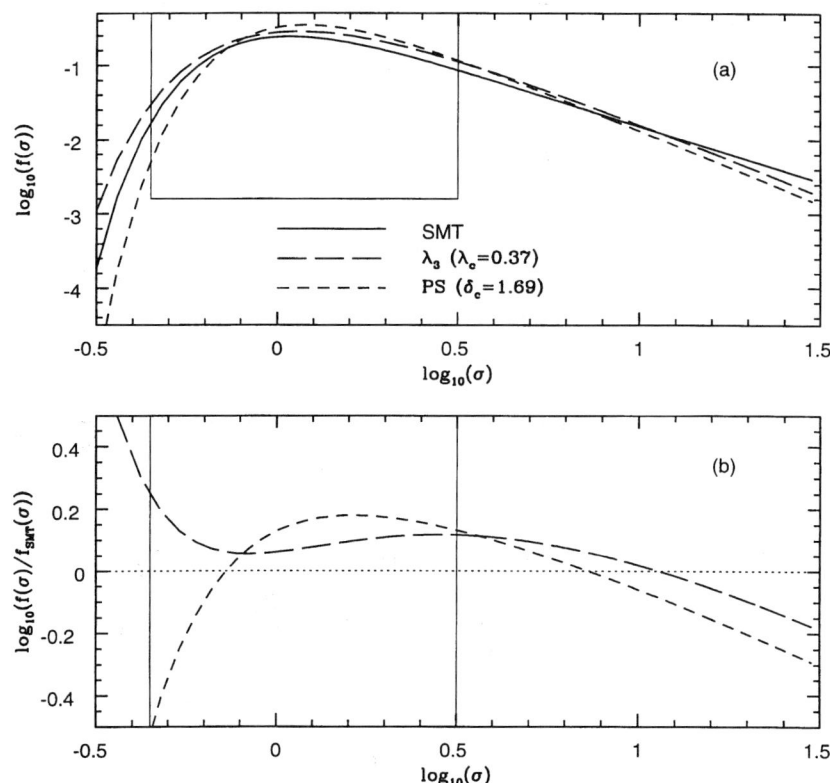

FIGURE 1. (a) The fraction of mass $f = dF/d\sigma$ in the gravitationally bound objects as a function of σ as predicted by the Press–Schechter model (*short dashed line*), Sheth–Mo–Tormen model (*solid line*), and λ_3-model (*long dashed line*). The small box shows the range of σ where the models have been checked against N-body simulations (see Figure 2 in [20]). (b) The logarithm of the ratios f_{PS}/f_{SMT} (*short-dashed line*) and f_{λ_3}/f_{SMT} (*long-dashed line*).

5. COMPARISON WITH N-BODY SIMULATIONS

FIGURE 2 shows the comparison of the λ_3-function with the numerical mass functions for the scale invariant initial spectra: $P(k) \propto k^n$ with $n = -1$ and $n = 0$ (for details see [30]). The top panel ($n = -1$) shows good agreement of the λ_3-function with the numerical mass function, while in the $n = 0$ case the agreement is significantly worse. FIGURE 3 compares the λ_3-function with the N-body simulation of the SCDM model [31]. At four epochs ($z = 1.86, 1.14, 0.43$, and 0) the λ_3-function is in better agreement than the Press–Schechter mass function. A similar result has been reported by Sheth and Tormen [19].

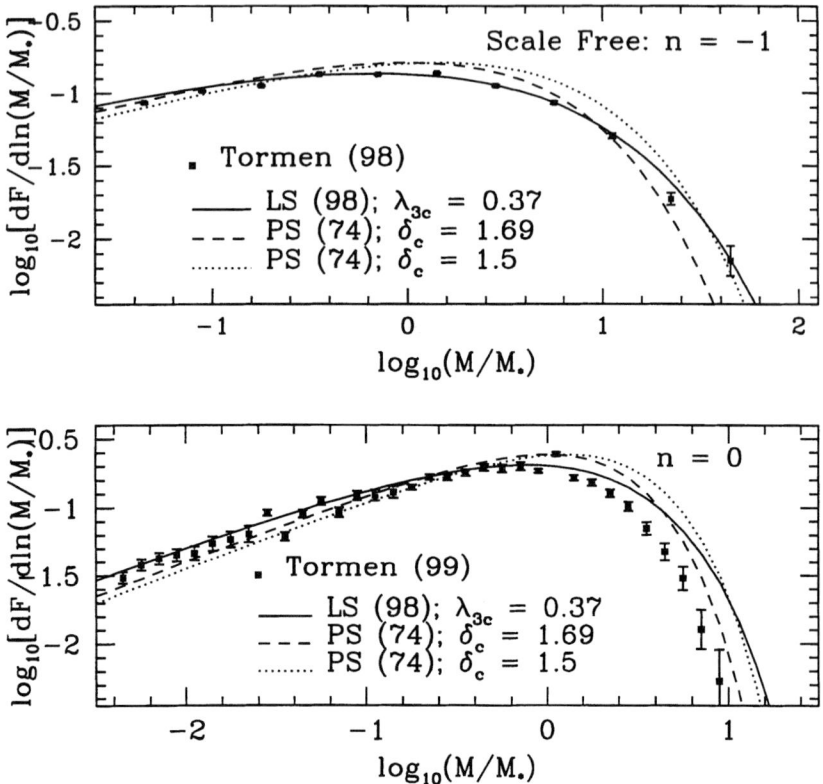

FIGURE 2. The *square dots* represent the numerical mass function with poissonian error bars. The *solid line* is the λ_3-mass function with $\lambda_{3c} = 0.37$ while the *dashed*, the *dotted* lines are the PS mass functions with $\delta_c = 1.69$, 1.5, respectively. The upper and the lower panels correspond to the $n = -1$ and the $n = 0$ power-law models, respectively. See also the top left panel of Figure 2 in [30].

6. LARGE-SCALE BIASING

It has been noticed for sometime that initial gravitational potential may noticeably affect large scale structure. Kofman and Shandarin [32] showed that the adhesion approximation predicts that the formation of voids is associated with positive peaks of primordial gravitational potential. Sahni *et al.* [33] studied the effect and measured a significant correlation between the sizes of voids and the value of primordial gravitational potential in numerical simulations of the adhesion model. Recently, Madsen *et al.* [34] demonstrated by N-body simulations that underdense and overdense regions are closely linked to regions with positive and negative gravitational potential, respectively. Lee and Shandarin [35] showed that initial potential also affects the masses of clusters.

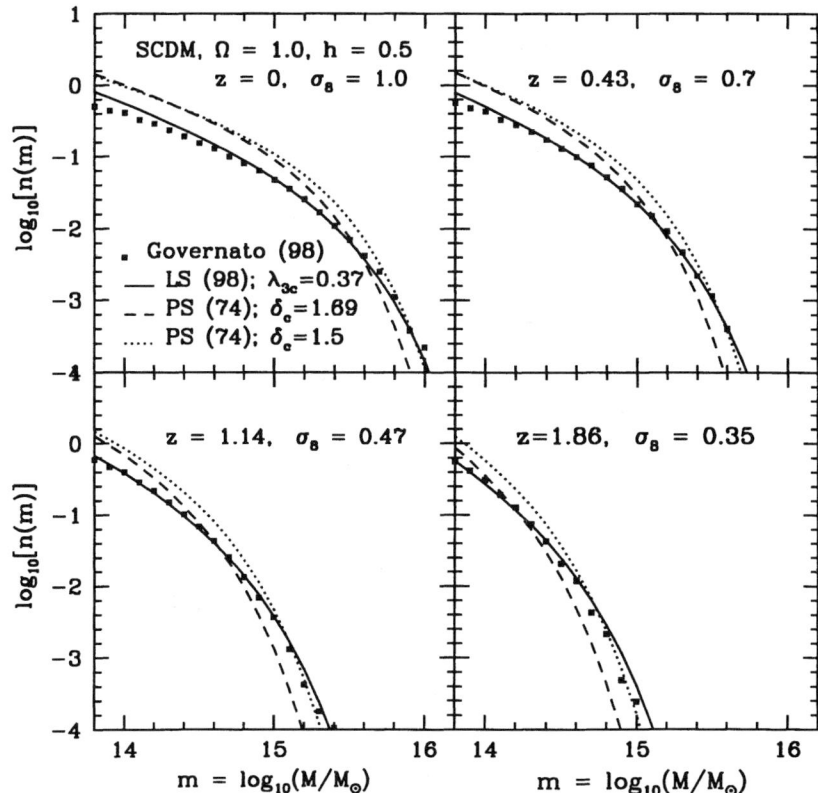

FIGURE 3. The *square dots* represent the numerical data for the case of SCDM model with $\Omega = 1$, $h = 0.5$. The *solid line* is our mass function with $\lambda_{3c} = 0.37$, and the *dashed*, the *dotted lines* are the PS mass functions with $\delta_c = 1.69$, 1.5, respectively.

In order to incorporate the primordial gravitational potential fluctuations term into the derivation of the mass function, we first derive the conditional probability density distribution $p(\delta|\varphi < -\varphi_c)$ $(\varphi_c > 0)$:

$$p(\delta|\varphi < -\varphi_c) = \frac{1}{\sqrt{2\pi}\sigma_\delta} \exp\left(-\frac{\delta^2}{2\sigma_\delta^2}\right)\left[1 - \mathrm{erf}\left(\frac{\varphi_c}{\sqrt{2}\sigma_\varphi}\right)\right]^{-1} \times$$

$$\left[1 + \mathrm{erf}\left(\frac{\kappa\dfrac{\delta}{\sigma_\delta} - \dfrac{\varphi_c}{\sigma_\varphi}}{\sqrt{2(1-\kappa^2)}}\right)\right] \quad (18)$$

Here σ_δ^2, σ_v^2, and σ_φ^2 are the density, velocity, and potential variances, respectively; $\kappa = <\delta \cdot \varphi>/\sigma_\delta\sigma_\varphi = \sigma_v^2/\sigma_\delta\sigma_\varphi$ is the crosscorrelation coefficient of the density con-

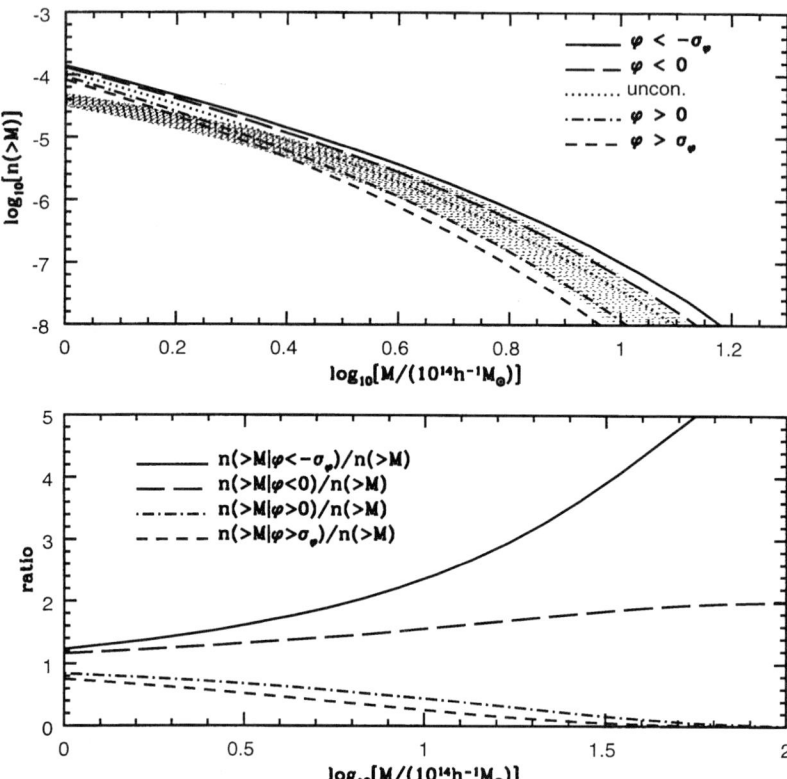

FIGURE 4. In the upper panel the conditional cumulative mass function satisfying chosen potential condition is plotted. The *solid*, the *long dashed*, the *dot-dashed*, and the *dashed lines* correspond to the conditions $\varphi < -\sigma_\varphi$, $\varphi < 0$, $\varphi > 0$, and $\varphi > \sigma_\varphi$, respectively, while the *dotted line* represents the unconditional cumulative PS mass function. The shaded area is 1σ fit to the observational cumulative mass function of rich clusters by Bahcall and Cen [5]. In the lower panel, the ratio of the conditional cumulative mass functions to the unconditional one is plotted for each condition. The CDM spectrum with $\Gamma = 0.25$ normalized to $\sigma_8 = 1$ has been used.

trast δ smoothed on the scale k_c and the primordial (*unsmoothed*) potential fluctuations φ. As a result, Eq. (1) for the conditional mass function $n(M|\varphi < -\varphi_c)$ becomes

$$n(M|\varphi < -\varphi_c) = -\frac{\bar{\rho}}{M}\left(\frac{\partial F}{\partial \sigma_\delta}\frac{d\sigma_\delta}{dM} + \frac{\partial F}{\partial \sigma_v}\frac{d\sigma_v}{dM}\right). \tag{19}$$

The further calculation needs to be done numerically. FIGURE 4 illustrates how the mass function depends on the initial potential in the CDM model with $\Gamma = \Omega h = 0.25$ normalized to $\sigma_8 = 1$. The top panel shows the mass function for regions of positive and negative initial potential as well as unconditional mass function. The bottom panel show the ratio of conditional mass functions to unconditional ones.

We also calculate the probability that a clump with mass M is located in the potential regions satisfying the chosen condition, for instance, $\varphi < -\varphi_c$

$$P(\varphi < -\varphi_c|M) = \frac{n(M|\varphi < -\varphi_c)}{n(M)} P(\varphi < -\varphi_c), \qquad (20)$$

where $P(\varphi < -\varphi_c)$ is the fraction of space satisfying the given condition (see FIG. 5). The scale of the initial potential

$$R_\varphi = \sqrt{3}\sigma_\varphi/\sigma_{\varphi'} = \sqrt{3\frac{\int_{k_l}^\infty dk k^{-2} P(k)}{\int_0^\infty dk P(k)}} \approx 120 h^{-1} \text{ Mpc} \qquad (21)$$

does not depend on any *ad hoc* scale; the dependence on k_l is extremely weak ($\propto \sqrt{\ln(1/k_l)}$ for the Harrison–Zel'dovich spectra assumed here). The geometry of gravitational potential does not evolve much on large scales [32], [34]. Therefore, the potential at present is very similar to the primordial one on scales much greater than the scale of nonlinearity. A simple explanation to this in the frame of the standard scenario of structure formation is due to the fact that the mass has been displaced by a distance of approximately $10\ h^{-1}$ Mpc [36]. Therefore, the potential on scales greater than, say, $30\ h^{-1}$ Mpc has been changed very little.

For the model in question the scale of the primordial potential is found to be $R_\varphi \approx 120\ h^{-1}$ Mpc. The scale of the density contrast field reaches this value $R_\delta = \sqrt{3}\ \sigma_\delta/\sigma_{\delta'} \approx 120\ h^{-1}$ Mpc only after it is smoothed on $k_c \approx 0.017\ h\text{Mpc}^{-1}$. The corresponding density variance on this scale is $\sigma_\delta(0.017\ h\text{Mpc}^{-1}) \approx 0.03$. On the other hand, the number of clumps with masses $10^{14} - 10^{15}\ h^{-1}\ M_\odot$ can easily be 30% greater in the troughs of the potential than the mean density $n(>M) = 0.5[n(>M|\varphi<0) + n(>M|\varphi > 0)]$ (see FIG. 4). Thus, the bias factor b (defined by the relation $\Delta n_{cl}/n_{cl} = b\ \Delta\rho_m/\rho_m$) reaches at least 10 on the scale about $120\ h^{-1}$ Mpc.

FIGURE 5 demonstrates that the most massive clusters ($M > 10^{15} h^{-1}\ M_\odot$) are almost certainly located in the the troughs in the initial potential. The bias defined as the density contrast of the clusters with respect to the mass density contrast $b = \delta_{cl}/\delta\rho$ reaches the value 3–10 on the scale of the potential $R_\phi \approx 120 h^{-1}$ Mpc [35].

SUMMARY

I discussed new modifications of the Press–Schechter theory of the cosmological mass function. One assumes a different collapse condition implying that a fluid particle becomes part of a gravitationally bound object after it experiences collapses along three axes. The comparison with other models (FIG. 1) shows that it predicts about 25% more gravitationally bound clumps than the Sheth–Mo–Tormen model in the range $0.45 \geq \sigma \geq 3.1$ where the comparison with the N-body simulations has been done. A direct comparison with N-body simulations (FIGS. 2 and 3) shows good agreement although not as good as the Sheth–Mo–Tormen model. The λ_3-function based on the Zel'dovich approximation has been obtained analytically similar to the

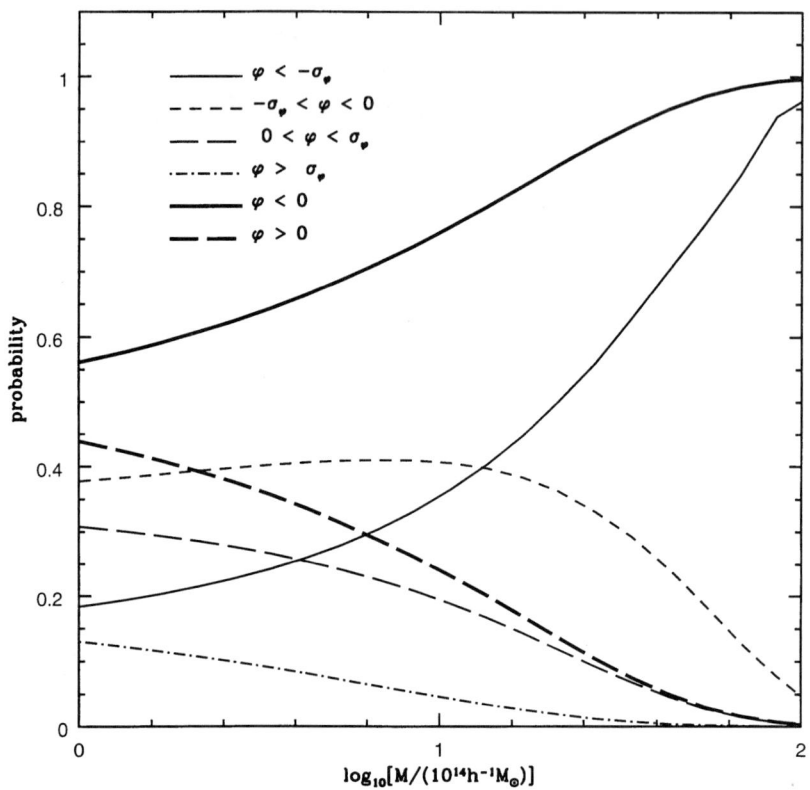

FIGURE 5. The probability that a clump with mass M can be found in the regions satisfying chosen potential condition is plotted. The *heavy solid*, the *heavy dashed*, the *solid*, the *dashed*, the *long dashed*, and the *dot-dashed lines* correspond to the condition $\varphi < 0$, $\varphi > 0$, $\varphi < -\sigma_\varphi$, $-\sigma_\varphi < \varphi < 0$, $0 < \varphi < \sigma_\varphi$, and $\varphi > \sigma_\varphi$, respectively.

Press–Schechter function. A drawback of the derivation is a crude approximation of the probability function $P(M,M_1)$ in the normalization integral equation (13). A more accurate normalization will be reported separately. The Sheth–Mo–Tormen model also suffers from a normalization problem: the shape of the mass function has been derived but the normalization has been enforced by demanding equality of Eq. (17).

Another modification is the conditional mass function showing that the clusters of galaxies tend to form in the troughs of initial gravitational potential and avoid the peaks of potential. The gravitational potential field has a typical scale of about $120h^{-1}$ Mpc and as a result has an advantage of being independent of the arbitrariness of the smoothing scale (if the filter scale is smaller than roughly $50h^{-1}$ Mpc) and at present it has almost same geometry as at the epoch of decoupling. Figures 4 and 5 quantify this large-scale biasing.

ACKNOWLEDGMENTS

I am grateful to Ravi Sheth for useful discussions during the workshop. This work was supported in part by the University of Kansas GRF 2001 grant.

REFERENCES

1. MONACO, P. 1998. Fundamentals of Cosmic Physics **19**: 157.
2. BOND, J.R. & S.T. MYERS. 1996. Ap.J.S. **103**: 1.
3. BAHCALL, N.A. & X. FAN. 1998. Astrophys. J. **504**: 1.
4. REICHART, D.E., R.C. NICHOL, F.J. CASTANDER, D.J.BURKE, A.K. ROMER, B.P. HOLDEN, C.A. COLLINS & M.P. ULMER. 1999. Astrophys. J. **518**: 521.
5. BAHCALL, N.A. & R. CEN. 1993. Astrophys. J. Lett. **407**: L49.
6. GIRARDI, M., S. BORGANI, R. GIURICIN, F. MARDIROSSIAN & M. MEZZETTI. 1998. Astrophys. J. **506**: 45.
7. REIPRICH, T. H. & H. BÖHRINGER. 1999. astro-ph/9908357.
8. PRESS, W. H. & P. SCHECHTER. 1974. Astrophys. J. **187**: 425.
9. PEACOCK, J. A. & A.F. HEAVENS. 1990. Mon. Not. R. Astr. Soc. **243**: 133.
10. BOND, J.R., S. COLE, G. EFSTATHIOU & N. KAISER. 1991. Astrophys. J. **379**: 440.
11. EFSTATHIOU, G., C.S. FRENK, S.D.M. WHITE & M. DAVIS. 1988. Mon. Not. R. Astr. Soc. **235**: 715.
12. CARLBERG, R.G. & H.M.P. COUCHMAN. 1989. Astrophys. J. **340**: 47.
13. EFSTATHIOU, G. & M.J. REES. 1988. Mon. Not. R. Astr. Soc. **230**: P5.
14. KLYPIN, A., S. BORGANI, J. HOLTZMAN & J. PRIMACK. 1995. Astrophys. J. **444**: 1.
15. SHAPIRO, P.R., I.T. ILIEV & A.C. RAGA. 1999. Mon. Not. R. Astr. Soc. **307**: 203.
16. MONACO, P. 1995, Astrophys. J. **447**: 23.
17. AUDIT, E., R. TEYSSIER & J-M. ALIMI. 1997. Astron. Astrophys. **325**: 439.
18. LEE, J. & S.F. SHANDARIN. 1998. Astrophys. J. **500**: 14.
19. SHETH, R.K. & G. TORMEN. 1999. Mon. Not. R. Astr. Soc. **308**: 119.
20. SHETH, R.K., H.J. MO & G. TORMEN. 1999. astro-ph/9907024
21. LACEY, C. & S. COLE. 1994. Mon. Not. R. Astr. Soc. **271**: 676.
22. JEDAMZIK, K. 1995. Astrophys. J. **448**: 1.
23. YANO, T., M. NAGASHIMA & N. GOUDA. 1996. Astrophys. J. **466**: 1.
24. ZEL'DOVICH, YA. B. 1970. Astron. Astrophys. **5**: 84.
25. SHANDARIN, S. F. & YA.B. ZEL'DOVICH. 1989. Rev. Mod. Phys. **61**: 185.
26. SHANDARIN, S. F. & A.A. KLYPIN. 1984. Sov. Astron. 28: 491.
27. ARNOL'D, V.I., S.F. SHANDARIN & YA.B. ZELDOVICH. 1982. Geophys. Astrophys. Fluid Dynamics **20**: 111.
28. DOROSHKEVICH, A.G. 1970. Astrofizika **6**: 581.
29. LEE, J. & S.F. SHANDARIN. 1999. Astrophys. J. Lett. **517**: 5.
30. GOVERNATO, F., A. BABUL, T. QUINN, P. TOZZI, C.M. BAUCH, N. KATZ & G. LAKE. 1999. Mon. Not. R. Astr. Soc. **307**: 949.
31. TORMEN, G. 1998. Mon. Not. R. Astr. Soc. **297**: 648.
32. KOFMAN, L. A. & S.F. SHANDARIN. 1988. Nature **334**: 129.
33. SAHNI, V., B.S. SATHYAPRAKASH & S.F. SHANDARIN. 1994. Astrophys. J. **431**: 20.
34. MADSEN, S., A.G. DOROSHKEVICH, S. GOTTLOBER & V. MULLER. 1998. Astron. Astrophys. **329**: 1.
35. LEE, J. & S.F. SHANDARIN. 1998. Astrophys. J. Lett. **505**: L75.
36. SHANDARIN, S.F. 1993. *In* Cosmic Velocity Fields, F.R. Bouchet & M. Lachieze-Rey, Eds.: p. 383. Editions Frontieres. Paris.

Lensing of the CMB: Non-Gaussian Aspects

MATIAS ZALDARRIAGA

*Institute for Advanced Study, School of Natural Sciences,
Princeton, New Jersey 08540 USA*

ABSTRACT: We compute the small angle limit of the three- and four-point function of the cosmic microwave background (CMB) temperature induced by the gravitational lensing effect by the large-scale structure of the universe. We relate the non-Gaussian aspects presented in this paper with those in our previous studies of the lensing effects. We interpret the statistics proposed in previous work in terms of different configurations of the four-point function and show how they relate to the statistic that maximizes the S/N.

KEYWORDS: Cosmic microwave background: cosmology; Theory: gravitational lensing

1. INTRODUCTION

The anisotropies in the cosmic microwave background (CMB) are thought to contain detailed information about the underlying cosmological model. In conventional models the anisotropies on most angular scales were created at the last scattering surface, at a redshift of $z \sim 1000$. At these early times the evolution of perturbations can be calculated accurately with linear theory. The calculation of theoretical predictions is almost straightforward, thus detailed observations of the microwave sky can, at least in principle, greatly constrain the cosmological model. We expect to be able to measure many of the parameters of the cosmological model with percent accuracy [1].

There are several physical processes that imprint anisotropies on the CMB after decoupling. Some of them will degrade our ability to learn about cosmology, like foreground emission from our galaxy. Others will allow us to constrain processes that happen after decoupling and help us understand the evolution of our universe. For example, the reionization of hydrogen by the ultraviolet light from the first generation of objects leaves a distinct mark in the polarization of the CMB [1], Sunyaev–Zeldovich emission from hot gas along the line of sight creates temperature anisotropies and the large-scale structure (LSS) of the universe deflects the CMB photons, lensing the anisotropies [2]–[7]. When studying the lensing effect produced by the large-scale structure of the universe we are trying to detect lensing produced by random mass fluctuations, the LSS, on a random background image, the CMB. The characteristics of the lensing effect depends on the relative size of the coherence lengths of these two random fields. In Seljak and Zaldarriaga [5] and Zaldarriaga and Seljak [6] we studied the limit of a rapidly fluctuating CMB background being lensed by a slowly varying mass distribution. This is the appropriate limit to recover

---

Address for corespondence: Matias Zaldarriaga, Department of Physics, New York University, 4 Washington Place, New York, NY 10003 USA. Voice: 212/998-7655; fax: 212/995-4016.
matiasz@physics.nyu.edu

the power spectrum of the projected mass density on scales much larger than the coherence length of the CMB, $\xi \sim 0.15^o$. In Zaldarriaga [8] we study the opposite limit, the generation of power on scales much smaller than ξ.

The lensing effect is expected to be the dominant nonprimordial contribution to the CMB anisotropies on small scales ($l \sim 3000$). It has been shown [9] that an accurate determination of the power generated by lensing can help break some of the parameter degeneracies in the CMB. Interferometric observations of the anisotropies such as those that will be carried out by CBI[1] are designed to make measurements at these angular scales.

To be able to use the observed power on small scales to break the degeneracies in the parameters one must be sure that one is observing the lensing signal. In Zaldarriaga [8] we showed that the small-scale power generated by gravitational lensing has a very definite signature, it is correlated with the large-scale gradient. Regions of the sky where the large-scale gradient is larger will have more small-scale power. The physical effect can be understood easily in the case of a cluster of galaxies lensing a smooth CMB gradient [10].

In this paper we will present the general three- and four-point function of the temperature field induced by lensing and show that both the statistic discussed in Zaldarriaga [8] and those proposed in Zaldarriaga and Seljak [6] are particular subsets of the possible configurations of the four-point function. We will present the statistic that maximizes the signal-to-noise ratio.

2. THE THREE- AND FOUR-POINT FUNCTIONS IN THE SMALL ANGLE LIMIT

In our previous studies [5], [8], [11], [12], we have investigated several ways of detecting the effect of gravitational lensing on the CMB. As we have explained, this amounts to trying to detect the distortions on the random CMB maps created by the random distributions of the dark matter in the universe. In [12] we used the ISW (integrated Sachs–Wolfe) effect as a tracer of the dark matter distribution and combinations of the CMB derivatives to measure the effect of lensing. The cross correlation of these two effects allowed us to gain information about the time evolution of the gravitational potential. Our method combined the information in particular configurations of the three-point function of the temperature. Other studies have used other combinations of the bispectrum to detect the signal [4] and also calculated the contributions to the bispectrum coming from other secondary procesess [4], [13].

In Zaldarriaga and Seljak [5] we used the power spectrum of quadratic combination of derivatives of the CMB to measure the power spectrum of the projected mass density κ. This method was valid in the limit in which we wanted to recover the long wavelength modes of κ from information in the small-scale CMB. This regime is analogous to weak lensing of background galaxies. In essence the different estimates of the power spectrum of κ at different scales were obtained by combining different configurations of the four-point function of the lensed temperature.

---

[1]Information on CBI can be found at http://astro.caltech.edu/~tjp/CBI/abstract.html.

In Zaldarriaga [8] we studied other configurations of the four-point function to illustrate the nature of the non-Gaussianities induced by lensing on small scales. The non-Gaussian nature of the generated power manifested itself in the correlations between the large-scale gradient and the small-scale generated power.

In order to have a unified picture of the different statistics we have proposed it is convenient to study directly the four-point function of the temperature field and a three-point function which correlates two temperatures and another field X. The field X stands for any field that cross correlates with κ. In Seljak and Zaldarriaga [12] $X = T$, but one can imagine doing this correlation with other tracers of the mass, like the fluctuations of the far infrared background.

We define the connected three- and four-point functions as,

$$\langle X(l_1)T(l_2)T(l_2)\rangle_c = (2\pi)^2 \delta^D(l_{123}) T_2(l_1,l_2,l_2)$$

$$\langle T(l_1)T(l_2)T(l_2)T(l_4)\rangle_c = (2\pi)^2 \delta^D(l_{1234}) T_4(l_1,l_2,l_2,l_4),$$

(1)

Gravitational lensing produces,

$$T_2(l_1,l_2,l_2) = 2 C_{l_1}^{\kappa X} \left[C_{l_2}^{\tilde{T}\tilde{T}} \frac{l_2 \cdot l_1}{l_1^2} + C_{l_3}^{\tilde{T}\tilde{T}} \frac{l_3 \cdot l_1}{l_1^2} \right]$$

$$T_4(l_1,l_2,l_2,l_4) = C_{l_1}^{\tilde{T}\tilde{T}} C_{l_2}^{\tilde{T}\tilde{T}} \left[\frac{(l_1+l_3)\cdot l_1 (l_1+l_3)\cdot l_2}{\|l_1+l_4\|^2} C_{l_{13}}^{\delta\delta} \right.$$

$$\left. + \frac{(l_1+l_4)\cdot l_1 (l_1+l_4)\cdot l_2}{\|l_1+l_4\|^2} C_{l_{14}}^{\delta\delta} \right] + \text{permutations}$$

(2)

(5 terms proportional to

$$C_{l_1}^{\tilde{T}\tilde{T}} C_{l_3}^{\tilde{T}\tilde{T}}, C_{l_1}^{\tilde{T}\tilde{T}} C_{l_4}^{\tilde{T}\tilde{T}}, C_{l_2}^{\tilde{T}\tilde{T}} C_{l_3}^{\tilde{T}\tilde{T}}, C_{l_2}^{\tilde{T}\tilde{T}} C_{l_4}^{\tilde{T}\tilde{T}}, C_{l_3}^{\tilde{T}\tilde{T}} C_{l_4}^{\tilde{T}\tilde{T}})$$

The unconnected part of the four-point function also gets corrections. To make the calculation of these terms fully consistent up to second order in the deflection angle we need to consider the contributions coming from the second order in the deflection angle as well. The unconnected terms are not relevant for our study so we will not write them down here.

In our previous papers we introduced three variables $\mathcal{E}$, $\mathcal{B}$, and $\mathcal{S}$. We had defined them in terms of derivatives to the temperature field. Equivalently we can write,

$$\mathcal{S}(l) = \int \frac{d^2 l_1}{(2\pi)^2} (l-l_1)\cdot l_1 T(l-l_1) T(l_1)$$

$$\mathcal{Q}(l) = \int \frac{d^2 l_1}{(2\pi)^2} [(l_x - l_{1x})l_{1x} - (l_y - l_{1y})l_{1y}] T(l-l_1) T(l_1)$$

$$\mathcal{U}(l) = \int \frac{d^2l_1}{(2\pi)^2} [(l_x - l_{1x})l_{1y} + (l_y - l_{1y})l_{1x}] \, T(l-l_1)T(l_1)$$

$$\mathcal{E}(l) = \mathcal{Q}(l)\cos(2\phi_l) + \mathcal{U}(l)\sin(2\phi_l)$$

$$\mathcal{B}(l) = -\mathcal{Q}(l)\sin(2\phi_l) + \mathcal{U}(l)\cos(2\phi_l) \tag{3}$$

When we average over the CMB random field we get,

$$\langle \mathcal{S}(l) \rangle_{CMB} = ((2\pi)^2 \delta^D(l) - 2\kappa(l))\sigma_S$$

$$\langle \mathcal{E}(l) \rangle_{CMB} = -2\kappa(l)\,\sigma_S \tag{4}$$

$$\langle \mathcal{B}(l) \rangle_{CMB} = 0.$$

We have introduced $\sigma_S = \int l\,dl/2\pi\, l^2 C_l^{\tilde{T}\tilde{T}}$.

To extract all the information in the three-point function we combine all possible configurations with a weight β chosen to maximize the signal to noise ratio. We define,

$$\hat{Y} = \frac{A_f}{(2\pi)^2} \int \frac{d^2l_1}{A_l} \frac{d^2l_2}{A_l} \beta(l_1,l_2,l_3)\,X(l_1)T(l_2)T(l_3). \tag{5}$$

For these mildly non-Gaussian maps, the variance can be calculated by only taking the Gaussian part of the temperature, so that

$$\text{Var}(\hat{Y}) = A_f (2\pi)^2 \int \frac{d^2l_1}{A_l^2} \frac{d^2l_2}{A_l^2} \beta^2(S)\, 2 C_{l_1}^{XX} C_{l_2}^{\tilde{T}\tilde{T}} C_{l_3}^{\tilde{T}\tilde{T}}. \tag{6}$$

The power spectra in Eq. (6) must include the contribution from detector noise. There are additional terms in the variance if the field X and T had some cross correlation before lensing. In practice these terms are unimportant if one is interested in measuring the cross correlation $C_l^{\kappa X}$ at large angular scales (low l), as was the case in our study in [12]. This is so because most the information of lensing is encoded in the high l modes of the temperature, so it is effectively as if the integral over l_2 in Eq. (5) is done over high l modes while the l_1 integral only involves low l. Thus the terms that would involve C_l^{TX} are absent because there are no pair of triangles in which X and T are evaluated on the same l.

The weight β that maximizes the S/N is $\beta \propto T_2(l_1,l_2,l_3)/2\, C_{l_1}^{XX} C_{l_2}^{\tilde{T}\tilde{T}} C_{l_3}^{\tilde{T}\tilde{T}}$. Finally we get,

$$\left(\frac{S}{N}\right)^2 = A_f^{-1} \int d^2l_1 \int \frac{d^2l_2}{(2\pi)^2} \frac{T_3^2}{2 C_{l_1}^{XX} C_{l_2}^{\tilde{T}\tilde{T}} C_{l_3}^{\tilde{T}\tilde{T}}}$$

$$= A_f^{-1} \int d^2l_1 \frac{4(C_l^{\kappa X})^2}{C_{l_1}^{XX} C_l^{\text{eff}}} \tag{7}$$

$$\frac{1}{C_l^{\text{eff}}} \equiv \int \frac{d^2l_2}{(2\pi)^2} \left[C_{l_2}^{\tilde{T}\tilde{T}} \frac{l_2 \cdot l_1}{l_1^2} + C_{l_2}^{\tilde{T}\tilde{T}} \frac{l_3 \cdot l_1}{l_1^2} \right]^2 \frac{1}{2 C^{\tilde{T}\tilde{T}}(l_2) C^{\tilde{T}\tilde{T}}(l_3)}.$$

The power spectra in the denominator include the contribution from detector noise (amplified by the beam response $C_l \to C_l + B^2 N_l$. The easiest way to calculate C_l^{eff} is to use a Monte Carlo technique. We used the implementation of the VEGAS algorithm in numerical recipes.

The above formula can be compared to the what we obtained using the S and $\mathcal{E}$ variables in our previous work. Equation (7) of Seljak and Zaldarriaga [12] reads,

$$\left(\frac{S}{N}\right)^2 = \int d^2 l_1 \frac{4(C_l^{\kappa X})^2}{C_l^{XX}} W^2(l) \left(\frac{1}{N_l^{SS}} + \frac{1}{N_l^{\mathcal{E}\mathcal{E}}}\right), \tag{8}$$

where $W_{(l)}^2$ is a window that encapsulates the effect of beam smearing. For low l N_l^{SS} and $N_l^{\mathcal{E}\mathcal{E}}$ are constant and satisfy, $N_l^{SS} = 2 N_l^{\mathcal{E}\mathcal{E}}$.

In FIGURE 1 we compare $1/C_l^{\text{eff}}$ to

$$W^2(l) \frac{1}{N_l^{SS}} + \frac{1}{N_l^{\mathcal{E}\mathcal{E}}}.$$

We show the results for two separate examples, an ideal experiment with no noise and infinite resolution and the Planck satellite. We focus on the large-scale limit and for the ideal experiment we only consider the information coming from modes with $l < 3000$. There are several salient features of the comparison. Although the difference between the methods is not so large for Planck it is much larger for the ideal experiment. This can be easily understood. In our previous method the power spectrum of the CMB noise in this limit was,

$$N_l^{SS} = (2\pi) \frac{\int l^5 dl (C_l^{\tilde{T}\tilde{T}})^2}{(\int l^3 dl C_l^{\tilde{T}\tilde{T}})^2}. \tag{9}$$

It is clear from Eq. (9) that once we get into the damping tail where the $C_l^{\tilde{T}\tilde{T}}$ fall exponentially, N_l^{SS} no longer changes which means that our method does not receive any information from those modes. In contrast Eq. (7) shows that the amplitude of the $C_l^{\tilde{T}\tilde{T}}$ cancels in C_l^{eff} as long as the modes had been measured with high S/N. Thus the new method continues to extract information from modes in the damping tail. In most practical cases this is not very important because the detector noise quickly dominates in this regime and then all methods downweight the modes. This explains why the difference between the two methods is not that large for Planck. There is another relevant consideration. When computing the variance we assumed that the field was only mildly non-Gaussian and that we could take the unlensed temperature power spectrum to calculate it. This is clearly not the case on small scales. As we have shown in previous sections, on small scales the fluctuations become very non-Gaussian as most of the power is generated by lensing. We only consider modes with $l < 3000$ for the calculation of C_l^{eff} to partially take into account this effect.

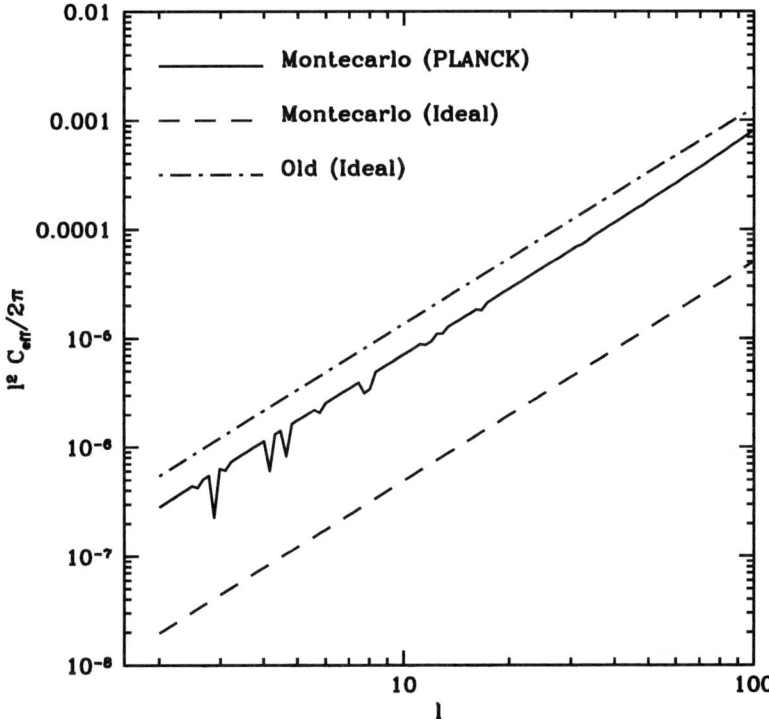

FIGURE 1. Comparison between C_1^{eff} and the noise of our old method. We show the results for the Planck satellite and an ideal experiment. For our old method there was no difference in the noise between Planck and an ideal experiment on these angular scales.

The Four-point Function

In Zaldarriaga [8] we studied one particular physical limit of the lensing effect, the recovery of information on the fluctuations of the mass distribution on scales much smaller than the coherence length of the CMB. We implemented a cross correlation between the large sclae gradients and the small-scale power that was sensitive to this effect. To recover this limit from our full four-point function expression we have to consider a quadrilateral in which two sides are much smaller that the other two (FIG. 2a). The two small sides correspond to the large-scale gradients while the large l correspond to the small-scale power. We consider the case where $l_1, l_2 \ll l_2 \sim l_4$. As we noted before the power spectrum of the primary anisotropies decreases exponentially while that of the deflection angle is only as a power law. We conclude that of all the terms in Eq. (2) only those explicitly written dominate,

FIGURE 2. Quadrilaterals corresponding to the two limits discussed in the text. **Panel a** corresponds to configurations relevant for the cross correlation between the large-scale gradient and the small-scale power. **Panel b** shows the configurations that enter in the calculation of $\mathcal{S}$ and $\mathcal{E}$.

$$T_4(l_1,l_2,l_2,l_4) \approx 2\frac{l_3 \cdot l_1 l_3 \cdot l_2}{l_3^2} C_{l_3}^{\delta\delta} C_{l_1}^{\tilde{T}\tilde{T}} C_{l_2}^{\tilde{T}\tilde{T}}, \quad l_1, l_2 \ll l_2 \sim l_4, \tag{10}$$

where we approximated $C_{l_3}^{\delta\delta} \approx C_{l_4}^{\delta\delta}$.

A different set of quadrilaterals dominate in the calculation of $\mathcal{E}$ and $\mathcal{S}\sharp$. Those variables extract information about the large-scale κ fluctuations from small angular scale fluctuations in the CMB. If we focus on modes of the temperature on scales larger that the damping tail $l^2 C_l^{\tilde{T}\tilde{T}}$ remains approximately constant while the power spectra of the deflection angle falls. Moreover, $\mathcal{S}$ and $\mathcal{E}$ are combination of derivatives of the CMB and the extra ls weigh the contribution to smaller scales. The power spectra of $\mathcal{S}$ and $\mathcal{E}$ are dominated by the type of quadrilaterals shown in FIGURE 2b. All the ls are large but the quadrilaterals are thin. The thin diagonal corresponds to the l of the κ mode being recovered. The terms in Eq. (2) proportional to $C_{l_1}^{\tilde{T}\tilde{T}} C_{l_2}^{\tilde{T}\tilde{T}} C_{l_{12}}^{\delta\delta}$ (where l_1 and l_2 represent the length of the sides and l_{12} is the small diagonal) dominate.

To extract all the information in the four-point function we can add all the quadrilaterals with an appropriate weight,

$$\hat{Z} = \frac{A_f}{(2\pi)^2} \int \frac{d^2l_1}{A_l} \frac{d^2l_2}{A_l} \frac{d^2l_3}{A_l} \beta(\mathcal{S}) T(l_1)T(l_2)T(l_2)T(l_4), \tag{11}$$

where $l_4 = -(l_1 + l_2 + l_2)$ and A_l is the area in l space we are using. For the optimal filter that minimizes S/N one gets

$$\beta \propto T_4 / C_{l1}^{\tilde{T}\tilde{T}} C_{l2}^{\tilde{T}\tilde{T}} C_{l3}^{\tilde{T}\tilde{T}} C_{l4}^{\tilde{T}\tilde{T}},$$

where we have assumed Gaussianity to compute the variance. For the S/N we get,

$$\left(\frac{S}{N}\right)^2 = \frac{A_f^{-1}}{24(2\pi)^4} \int d^2l_1\, d^2l_2\, d^2l_3 \frac{T_4^2(l_1, l_2, l_3, l_4)}{C_{l1}^{\tilde{T}\tilde{T}} C_{l2}^{\tilde{T}\tilde{T}} C_{l3}^{\tilde{T}\tilde{T}} C_{l4}^{\tilde{T}\tilde{T}}} \tag{12}$$

where the power spectra in the denominator should include the contribution form detector noise.

We change integration variables in Eq. (12) and write,

$$\left(\frac{S}{N}\right)^2 = A_f^{-1} \int d^2 l_1 \left(\frac{4 C_{l_1}^{\kappa\kappa}}{\tilde{C}_l^{\text{eff}}}\right)^2$$

$$\left(\frac{1}{\tilde{C}_l^{\text{eff}}}\right)^2 = \frac{1}{24} \int \frac{d^2 l_2}{(2\pi)^2} \frac{d^2 l_3}{(2\pi)^2} \frac{T_4^2(l_1 - l_2, l_2, l_3, -l_1 - l_3)}{C_{l_1 - l_2}^{\tilde{T}\tilde{T}} C_{l_2}^{\tilde{T}\tilde{T}} C_{l_3}^{\tilde{T}\tilde{T}} C_{l_1 + l_3}^{\tilde{T}\tilde{T}} (4 C_{l_1}^{\kappa\kappa})^2}$$

(13)

This is a useful change of variables because it makes the integral resemble what we had in our old method. In this way the limit $l_1 \to 0$ corresponds to quadrilaterals that have four large sides but a small diagonal (l_1). Equation (6) of Seljak and Zaldarriaga [11] reads,

$$\left(\frac{S}{N}\right)^2 = A_f^{-1} \int d^2 l_1 \frac{(4 C_l^{\kappa\kappa})^2}{\sigma_{C_l}^2},$$

(14)

with $\sigma_{C_l}^{-2} = \sigma_{C_l^{SS}}^{-2} + \sigma_{C_l^{\mathcal{TE}}}^{-2} + \sigma_{C_l^{S\mathcal{E}}}^{-2}$.

In FIGURE 3 we compare the results C_l^{eff} in Eq. (13) with σ_{C_l} in Eq. (14). The plot is qualitatively similar to FIGURE 1 for the three-point function. While there are hardly any improvements in our previous method when we go from Planck to an ideal experiment, there are significant differences for the optimal filter which continues to gather information from the damping tail. For Planck the situation is different, both the optimal method and our old method obtain a similar amount of information from the data. Even though the optimal method is able to get information from the damping tail in the ideal case, this is unimportant for Planck because the finite size of the beam makes it impossible. The fact that our previous method seems to have slightly less noise than the optimal method when l is a few hundred is most probably an artifact. The noise in our old method was calculated using a Gaussian approximation which we had seen breaking down slightly in our simulations [6]. As for the three-point function we only included modes of the temperature with $l < 3000$ as the power generated by lensing is non-Gaussian so our estimate of the variance of Z is not valid on smaller scales.

In the limit $l_1 \to 0$, the four-point function becomes approximately,

$$T_4(l_1 - l_2, l_2, l_3, -l_1 - l_2) \approx -C_{l_2}^{\tilde{T}\tilde{T}} C_{l_3}^{\tilde{T}\tilde{T}} l_1^2 C_{l_1}^{\delta\delta}.$$

(15)

To obtain Eq. (15) we had to assume that $C_l^{\kappa\kappa}$ is a decreasing function of l and that

$$C_{l_1 - l_2}^{\tilde{T}\tilde{T}} \approx C_{l_2}^{\tilde{T}\tilde{T}}$$

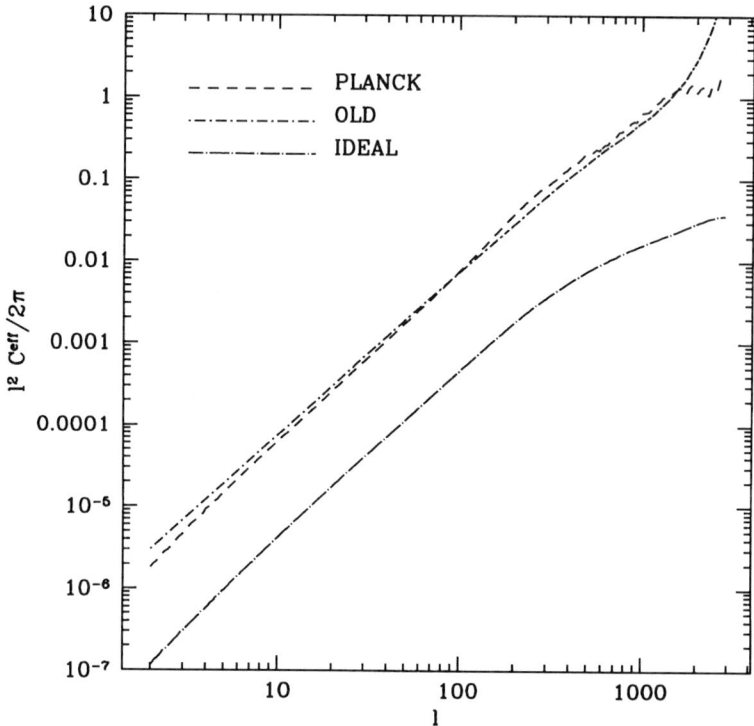

FIGURE 3. Result of Monte Carlo for four-point function together with result from previous technique.

and the equivalent formula for l_3. Both of these assumptions break down in some range of ls. For example, $C_l^{\kappa\kappa}$ has a peak at $l \sim 100$ and when l_2 is in the damping tale range, for finite l_1 there might be corrections due to the difference between

$$C_{l1-l2}^{\tilde{T}\tilde{T}} \text{ and } C_{l2}^{\tilde{T}\tilde{T}}.$$

We can still use this expression as a rough estimate to try to compare how the optimal β compares with the weight used by our previous method. In this limit and for the quadrilaterals relevant for this statistic, the optimal β is equivalent to multiplying each of the temperatures by

$$\frac{(C_l^{\tilde{T}\tilde{T}})^{1/2}}{C_l^{\tilde{T}\tilde{T}} + B_l^2 N_l^{\tilde{T}\tilde{T}}}.$$

Thus for a temperature power spectra that goes as $C_l^{\tilde{T}\tilde{T}} \propto l^{-2}$ and when detector noise is irrelevant, the optimal filter amounts to multiplying the temperatures by l, equivalent to taking derivatives. This is the reason our previous method is not far from optimal in situations where we can neglect detector noise and we are not trying to extract information form the damping tail of the CMB. As we mention when we discussed the three-point function, on small enough scales our treatment of the noise breaks down because the power is dominated by the power generated by lensing. The Gaussian approximation for the noise will not be valid.

3. CONCLUSIONS

We have presented the three- and four-point function of the lensed CMB field in the small angle limit. The different statistics introduced in our previous work to study the signatures of lensing on different angular scales are particular combinations of the four-point function of the temperature field. We have calculated explicitly the dependence of the three- and four-point functions on the CMB and deflection angle power spectra as well as on the shape of the configuration. The statistic introduced in previous papers can be viewed as particular ways of compressing the information in the four-point function that take into account the physical intuition coming from our understanding of the lensing effect. The lensing effect predicts a particular dependence of the four-point function on configuration and scale that can be used to separate it from other non-Gaussian signals.

ACKNOWLEDGMENTS

Matias Zaldarriaga is supported by NASA through Hubble Fellowship grant HF-01116.01-98A from STScI, operated by AURA, Inc. under NASA contract NAS5-26555.

REFERENCES

1. ZALDARRIAGA, M. 1997. Phys. Rev. D **55:** 1822.
2. BERNARDEU, F. 1997. Astron. & Astrophys. **432:** 15.
3. BERNARDEU, F. 1998. Astron. & Astrophys. **338:** 767.
4. GOLDBERG, D.M. & D.N. SPERGEL.1998. astro-ph/9811251.
5. SELJAK, U. & M. ZALDARRIAGA. 1999. Phys. Rev. Lett. **82:** 2636.
6. ZALDARRIAGA, M. & U. SELJAK. 1999. Phys. Rev. D **59:** 123507.
7. SELJAK, U. 1996. Astrophys. J. **463:** 1.
8. ZALDARRIAGA, M. 2000. Phys. Rev. D **62:** 063510.
9. METCALF, R.B. & J. SILK. 1998. Astrophys. J. Lett. **492:** L1.
10. SELJAK, U. & M. ZALDARRIAGA. 2000. Astrophys. J. **538:** 57.
11. SELJAK, U. & M. ZALDARRIAGA. 1999. Phys. Rev. Lett. **82:** 2636.
12. SELJAK, U. & ZALDARRIAGA M. 1999. Phys. Rev. D **60:** 043504.
13. COORAY, A. & W. HU. 2000. Astrophys. J. **534:** 533.

Cosmic Statistics of Statistics: N-point Correlations

ISTVÁN SZAPUDI

Institute for Astronomy, University of Hawaii, 2680 Woodlawn Drive, Honolulu, Hawaii 96822, USA

ABSTRACT: The fully general calculation of the cosmic error on N-point correlation functions and related quantities is presented. More precisely, the variance caused by the finite volume, discreteness, and edge effects is determined for *any* estimator which is based on a general function of N-tuples, such as multi-point correlation functions and multi-spectra. The results are printed explicitly for the two-point correlation function (or power-spectrum), and for the three-point correlation (or bispectrum). These are the most popular statistics in the study of large-scale structure, yet the general calculation of their variance has not been performed until now.

KEYWORDS: **Large-scale structure of the Universe; Methods: numerical**

1. INTRODUCTION

Astrophysics provides prime examples of random fields, such as the distribution of galaxies and the fluctuations of the cosmic microwave background (CMB). Such random fields can be characterized statistically by a series of well chosen estimators. The most popular ones include counts-in-cells, N-point correlation functions, as well as statistics derived from them. Indeed, there are mathematical theorems that state that under certain conditions, both series describe a random process fully.

Our goal in astrophysics is not simply to estimate these statistics, but to constrain underlying theories. This aim necessitates a firm control over the expected statistical errors from a survey. The theory of errors for finite surveys, the "cosmic statistics of statistics," is therefore of utmost importance for practical applications. Such a theory was formulated in the past for counts in cells statistics [9], [3], [2].

For the N-point correlation functions, however, to date only partial results are published, such as calculation of the discreteness effects for the two-point correlation function [7], and for N-point correlation functions for Poisson and multinomial point processes [10], full calculation for the two-point function under the hierarchical assumption with edge effects neglected [1], and some results in Fourier space with various degree of approximations.

The aim of this proceedings is to present the general variance calculation for N-point correlation functions with all major contributions included, such as discrete-

---

Address for correspondence: Istvan Szapudi, Institute for Astronomy, University of Hawaii, 2680 Woodlawn Drive, Honolulu, HI 96822 USA. Voice: 808/956-6196; fax: 808/956-9590.
szapudi@ifa.hawaii.edu

ness effects, arising from sampling the underlying random field with a finite number of galaxies, edge effects, due to the geometry of the survey and the corresponding uneven weighting of N-tuples, and finite volume effects, caused by fluctuations at and above the scale of a survey. The underlying technique of calculation, as well as the fully general results are presented here; specializations such as power spectrum and bispectrum, and approximations, such as weakly nonlinear perturbation theory and hierarchical assumptions, will be presented elsewhere in collaboration with Colombi and Szalay [11].

The next section sets up the general mathematical framework for the calculation using computer algebra packages. Section 3 presents the results for $N = 2$, and $N = 3$. The final discussion section outlines how the formulae can be used in practice, as well as describes developments in the immediate future.

2. GENERAL FRAMEWORK

Many interesting statistics, such as the N-point correlation functions and their Fourier analogs, can be formulated as functions over N points in a catalog. The covariance of a pair of such estimators will be calculated for a general point process under the assumption that the average density is *a priori* known. This is the obvious generalization of the Poisson process when arbitrarily high-order correlations are present. The number of objects is thus varied corresponding to a grand canonical ensemble in statistical physics. The following calculations lean heavily on the elegant formalism by Ripley [8], which can be consulted for details, and are the direct generalization of the framework set up by Szapudi and Szalay [10].

Let D be a catalog of data points to be analyzed, and R randomly generated over the same area, with averages λ, and ρ, respectively. The role of R is to perform a Monte Carlo integration compensating for edge effects, and therefore eventually the limit $\rho \to \infty$ will be taken. λ on the other hand is assumed to be externally estimated with arbitrary precision. No other assumption is taken about the point process. In practice, λ is usually estimated from the same survey, which gives rise to additional correlations, the "integral constraint bias." This effect will be investigated in more detail elsewhere.

Following [10], let us define symbolically an estimator $D^p R^q$, with $p + q = N$ for a function Φ symmetric in its arguments

$$D^p R^q = \Sigma \Phi(x_1, \ldots, x_p, y_1, \ldots, y_q), \qquad (1)$$

with $x_i \neq x_j \in D$, $y_i \neq y_j \in R$. As an example, the two-point correlation function corresponds to $\Phi(x,y) = [x,y \in D, r \leq d(x,y) \leq r + dr]$, where $d(x,y)$ is the distance between the two points, and [*condition*] equals 1 when *condition* holds, 0 otherwise. Ensemble averages can be estimated via factorial moment measures, ν_s [5], [8]. In the Poisson limit $\nu_s = \lambda^s \mu_s$, where μ_s is the s dimensional Lebesgue measure, and in the most general case $\nu_s f(x_1, \ldots, x_s) \lambda^s \mu_s$. The function $\lambda^s f(x_1, \ldots, x_s) = F(x_1, \ldots, x_s)$ can be identified as the full, that is, nonreduced, N-point correlation function. The connection between these and the reduced N-point correlation functions is well known [10], and can be obtained through either generating functions, or recursions.

The general covariance of a pair of estimators is

$$E(p_1, p_2, N_1, N_2) =$$

$$\langle \hat{D}_a^{p_1} \hat{R}_a^{q_1} \hat{D}_b^{p_2} \hat{R}_b^{q_2} \rangle = \sum_{i,j} \binom{p_1}{i} \binom{p_2}{i} i! \binom{q_1}{j} \binom{q_2}{j} j! S_{i+j} \lambda^{-i} \rho^{-j}, \quad (2)$$

with $p_1 + q_1 = N_1$, $p_2 + q_2 = N_2$, S will be specified later, $\wedge$ denotes normalization with λ, ρ, respectively, that is, $(\hat{D} = D/\lambda$, $\hat{R} = R/\rho$. The expression simply describes the fact that out of the p_1 and p_2 different data points in D we have an i-fold degeneracy, as well as a j-fold degeneracy in the random points drawn from R. To simplify the exposition of the calculation, it is convenient to assume from the very beginning the $\rho \to \infty$, that is, the random process employed for the Monte Carlo integration of the shape of the survey is arbitrarily dense. This is usually achievable in practice, thus it will not be considered further. Equation (2) describes the cross-correlation of two estimators even for two different objects as well, for example, two particular bins of the two- and three-point correlation functions. An interesting special case, $N_1 = 1$ (the average density in the survey) and $N_2 \geq 2$, is needed for calculating the "integral constraint" correction.

When the random process is arbitrarily dense only $j = 0$ survives,

$$E(p_1, p_2, N_1, N_2) =$$

$$\langle \hat{D}_a^{p_1} \hat{R}_a^{q_1} \hat{D}_b^{p_2} \hat{R}_b^{q_2} \rangle = \quad (3)$$

$$\sum_i \binom{p_1}{i} \binom{p_2}{i} i! \lambda^{-i} \hat{S}_i f(1, 2, \ldots, p_1, N_1 + 1, \ldots, N_1 + p_2 - i),$$

where $\hat{S}$ is now an operator acting on f,

$$\hat{S}_k = \int \Phi_a(1 \ldots N_1) \, \Phi_b(1 \ldots i, N_1 + 1, \ldots, N_1 + N_{2-i}) \ldots \mu_{2N-k}. \quad (4)$$

The operator $\hat{S}_k$ is analogous to the phase space integral S_k [10]. The dot emphasizes that the integral can be performed only after $\hat{S}_k$ is acted on f which is part of the measure. The phase space has to be calculated in the general case via the full factorial moment measure of which f is an integral part. Throughout the paper we use the convention that $\binom{k}{l}$ is nonzero only for $k \geq 0$, $l \geq 0$, and $k \geq l$, and the variables x_i are denoted with i for simplicity. Here Φ_a and Φ_b denote two different functions, for instance corresponding to two radial bins of two estimators of the same or different orders. In formula (4) the symmetry of Φ in its arguments was heavily relied on to achieve the above "standard" representation of the operator.

The dependence of S_k on a, b, and N is not noted for convenience [10], but they will be assumed throughout the paper. The estimator [10] for the generalized N-point correlation function is $(\hat{D} - \hat{R})^N$, or more precisely,

$$\tilde{w}_N = \frac{1}{S}\sum_i \binom{N}{i}(-)^{N-i}\left(\frac{D}{\lambda}\right)^i\left(\frac{R}{\rho}\right)^{N-i},\tag{5}$$

where $S = \int\Phi\mu_N$ (without subscript). In this case S is a number since it corresponds to the Poisson catalog with its simple factorial moment measure. The average of the estimator yields

$$\begin{aligned}\langle\tilde{w}_N\rangle &= \frac{1}{S}\sum_i\binom{N}{i}(-)^{N-i}\int\Phi(1,...,N)f(1,...,N)\mu_N,\\ &= \frac{1}{S}\int\Phi(1,...,N)\xi_N(1,...,N)\mu_N\end{aligned}\tag{6}$$

Since the role of Φ is effectively a window, with a window operator $\hat{W}$ this can be written symbolically as $\langle\tilde{w}_N\rangle = \hat{W}\xi_N/\hat{W}$. The asymptotic centered covariance between two estimators of the above for a general point process in the limit of $\rho \to \infty$ can be written according to the previous considerations as

$$\begin{aligned}\langle\delta\tilde{w}_{N_1}\delta\tilde{w}_{N_2}\rangle &= \langle\tilde{w}_{a,N_1}\tilde{w}_{b,N_2}\rangle - \langle\tilde{w}_{a,N_1}\rangle\langle\tilde{w}_{b,N_2}\rangle =\\ &\frac{1}{S^2}\sum_i\binom{N_1}{i}\binom{N_2}{j}(-)^{i+j}[E(i,j,N_1,N_2) - \hat{S}_0 f(1,...,i)f(N_1+1,...,N_1+j)],\end{aligned}\tag{7}$$

In Eq. (7) the operator S_0 acts on the multiple of the two f's on the right. Equation (7) is the main result of this paper. While it is quite cumbersome, it is easily expandable with the help of computer algebra, as demonstrated by the examples of the next section. The special cases rendered will also illustrate how the simplicity of the proposed class of estimators exactly manifests itself by a mass cancellation of terms. Any other estimator would have extra terms in the variance [10].

3. THE COSMIC ERROR ON THE TWO- AND THREE-POINT CORRELATION FUNCTIONS

Equation (7) was entered into a computer algebra package. The symmetries and simplicity of the estimator give rise to cancellations and possibilities for collecting similar terms. This is the reason why the final result for the two-point correlation function has only three to six terms, while formally it could have up to about a hundred. Alternative estimators, such as $DD/RR - 1$, etc., would not yield significantly less cancellations, therefore error-calculation for them was not attempted; although the same formalism applies.

3.1. The Two-point Function

The covariance for the two-point function (or any quantity estimated from doublets of points, such as the power spectrum) can be expressed after the cancellations and the possible simplifications as

$$\langle \delta \tilde{w}_2^a \delta \tilde{w}_2^b \rangle = \frac{1}{S_2} \Bigg\{ \int \Phi_a(1,2)\, \Phi_b(3,4)[\xi_4(1,2,3,4) + 2\xi(1,3)\xi(2,4)] +$$

$$\frac{4}{\lambda} \int \Phi_a(1,2)\, \Phi_b(1,3)[\xi(2,3) + \xi_3(1,2,3)] + \quad (8)$$

$$\frac{2}{\lambda^2} \int \Phi_a(1,2)\, \Phi_b(1,2)[1 + \xi(1,2)] \Bigg\}$$

Equation (8) is essentially identical to the result of Hamilton [6] where he calculates the variance of δ, the fluctuation field itself. This is not at all surprising [10]. The estimator contains exactly the same terms and coefficients (regardless of the choice of Φ) as δ itself, which strongly suggests that it is (nearly) optimal.

The above formula contains the different contributions to the error [9] entirely mixed. Approximate separation of the different terms appears to be fruitless. The only general point to be made is that discreteness effects are absent in the first term, while they are present (mixed with the other two effects) in the $1/\lambda^s$ terms, with $s > 0$. This is naturally true for the higher-order calculations as well.

It is worth reemphasizing that this formula applies for the generalized 2-point function, including the "traditional" 2-point function, and any of its incarnations, such as the power spectrum. In the latter case, esthetic or practical reasons might dictate that the errors on the power spectrum are expressed in terms of the power-spectrum, bi-, and tri-spectrum, instead of the two-, three-, and four-point correlation functions. Since the connection is a simple Fourier transform, this trivial exercise is left for the reader. Explicit formulae, aimed at practical applications for power-spectrum will be presented elsewhere [11].

3.2. The Three-point Correlation Function

The same method yields (co)variance for higher-order estimators as well. We present another example, the generalized three-point correlation function. Its variance, using again the main result of the proceeding, translates into:

$$\langle \delta \tilde{w}_3^a \delta \tilde{w}_3^b \rangle = \frac{1}{S^2} \{ \int \Phi_a(1,2,3)\Phi_b(4,5,6)[\xi(1,2,3,4,5,6) +$$
$$3\xi(1,2)\xi(3,4,5,6) + 9\xi(1,4)\xi(2,3,5,6) +$$
$$3\xi(4,5)\xi(1,2,3,6) + 9\xi(1,5,6)\xi(2,3,4) +$$
$$9\xi(1,4)\xi(2,3)\xi(5,6) + 6\xi(1,4)\xi(2,5)\xi(3,6)]$$
$$\frac{9}{\lambda} \int \Phi_a(1,2,3)\Phi_b(1,4,5)[\xi(1,2,3,4,5) +$$
$$\xi(2,3,4,5) + 2\xi(1,2)\xi(3,4,5) +$$
$$2\xi(1,4)\xi(2,3,5) + \xi(2,3)\xi(1,4,5) + \quad (9)$$
$$4\xi(2,5)\xi(1,3,4) + \xi(4,5)\xi(1,2,3) +$$
$$\xi(2,3)\xi(4,5) + 2\xi(2,4)\xi(3,5)] +$$
$$\frac{18}{\lambda^2} \int \Phi_a(1,2,3)\Phi_b(1,2,4)[\xi(1,2,3,4) +$$
$$2\xi(1,3,4) + \xi(1,2)\xi(3,4) +$$
$$2\xi(1,3)\xi(2,4) + \xi(3,4)]$$
$$\frac{6}{\lambda^3} \int \Phi_a(1,2,3)\Phi_b(1,2,3)[\xi(1,2,3) + 3\xi(1,2) + 1] \}.$$

For simplicity, in formula (9) the order of ξ is notated with the number of arguments only, for example, $\xi_3(1,2,3) = \xi(1,2,3)$. The above equation is less obviously useful then that of the two-point correlation function. Nevertheless, given a model for the higher-order correlation functions, such as weakly nonlinear perturbation theory, or any well-specified version of the hierarchical assumption, the equation can easily be turned into a practical computer program [11].

The variance of the four-point and higher-order correlation functions can be calculated analogously, but it would make no sense to print the results. When needed, the formulae generated by computer algebra can be transformed into Fortran or C-code directly.

4. DISCUSSION

The above method, and the explicit formulae given, can be used to evaluate the cosmic error on any statistical measure based on N-tuples of a distribution. This includes, but is not limited to, N-point correlation functions, Nth order cumulants, cumulant correlators, multispectra, etc.

The above calculation was performed only for the best N-point estimator [10]. Any other estimator can be calculated analogously, but be warned that the resulting number of terms can be overwhelming due to the insufficient cancellation arising from suboptimal edge correction.

The fact that the average density is assumed to be given as an outside parameter appears to be fairly restrictive. However, using maximum likelihood context, which is the most important potential practical application of the results, it is easy to remedy the situation. The proposed estimator [10] can be trivially changed by *not* normalizing with the average density λ. This introduces only a small modification to the final formulae, and a set of estimators, including that of the average count, contains all information needed for constructing the likelihood function. Such a procedure yields full statistical description, takes into account fluctuations in the mean and the fact the average is estimated from the same surveys ("integral constraint"). Practical demonstration of this procedure will be presented elsewhere [11].

The proposed estimator used here is not connected for $N \geq 4$ [10]. Therefore, the calculations for the higher-order connected estimator should be modified for accurate results for the connected moments. This trivial but tedious generalization is left for future research.

The explicit formulae can be specialized for several cases, which will be presented elsewhere [11]. The interesting limits include Poisson, gaussian, weakly-nonlinear, strong correlations, hierarchy, shot noise limited, continuous limit, etc. The detailed discussion of these limits, and specializations to particular statistics, such as N--point correlation functions, multispectra, would go beyond the scope of the present exposition. Similarly, the main equation yields cross correlations between different statistics as well, a must for any investigation in the maximum likelihood framework.

Finally, it is worth noting that recent advances in algorithms for calculating N-point correlation functions render these objects more interesting then ever. Fast algorithms [4] will make it possible to calculate N-point functions from very large catalogs, be it artificial or real data, which undoubtedly will culminate in new insights into the subject. The formulae presented in this proceedings will provide the firm theoretical grounding for any such investigation.

ACKNOWLEDGMENTS

This work will form the integral part of a project in collaboration with S. Colombi and A.S. Szalay soon to be published [11].

REFERENCES

1. BERNSTEIN, G.M. 1994. The variance of correlation function estimates. Astrophys. J. **424**: 569.
2. BOSCHAN, P., I. SZAPUDI & A.S. SZALAY. 1994. On the accurate determination of the clustering hierarchy of galaxies. ApJS **93**: 65.
3. COLOMBI S., I. SZAPUDI & A.S. SZALAY. 1998. Effects of sampling on statistics of large-scale structure. Mon. Not. R. Astr. Soc. **296**: 253.
4. CONNOLLY, A.J., R.C. NICHOL, A. MOORE & I. SZAPUDI. 2000. In preparation.
5. DALEY, D.J. & D. VERE-JONES. 1972. *In* Stochastic Point Processes, P.A.W. Lewis, Ed. Wiley. New York.
6. HAMILTON, A.J.S. 1993. Toward better ways to measure the galaxy correlation function. Astrophys. J. **417**: 19.
7. LANDY, S.D. & A. SZALAY. 1993. Bias and variance of angular correlation functions. Astrophys. J. **412**: 64.
8. RIPLEY, B.D. 1988. Statistical Inference for Spatial Processes. Cambridge University Press. Cambridge.
9. SZAPUDI, I. & S. COLOMBI. 1996. Cosmic error and statistics of large-scale structure. Astrophys. J. **470**: 131.
10. SZAPUDI, I. & A.S. SZALAY. 1998. A new class of estimators for the N-point correlations. Astrophys. J. Lett. **494**: L41.
11. SZAPUDI, I., S. COLOMBI & A.S. SZALAY. 2000. In preparation.

APPENDIX: AN ALTERNATIVE TECHNIQUE

An alternative method of calculation is possible, which is instructive and insightful, even if less rigorous mathematically then the above formalism using factorial moment measures. This alternative technique is well suited for obtaining quick results for low order estimators by hand. We demonstrate the calculation for two-point correlation function, higher order results can be obtained analogously, although it quickly becomes prohibitevely tedious.

Let us assume that the survey is divided into K bins, each of them with fluctuations δ_i, with i running from $1\ldots K$. For this configuration our estimator can be written as

$$\tilde{w} = f_{12}\delta_1\delta_2. \qquad (11)$$

Equation (11) uses a "shorthand" Einstein convention: 1,2 substituting for i_1, i_2, and repeated indices summed. The pairwise weights f_{12} correspond to Φ in the main body of the paper, and it is assumed that the two indices cannot overlap.

The ensemble average of the above estimator is clearly $f_{12}\xi_{12}$. The continuum limit (co)variance between bins a and b can be calculated by taking the square of the above, and taking the ensemble average.

$$\langle \delta\tilde{w}^a \delta\tilde{w}^b \rangle = f_{12}^a f_{34}^b (\langle \delta_1\delta_2\delta_3\delta_4 \rangle - \langle \delta_1\delta_2 \rangle\langle \delta_3\delta_4 \rangle). \tag{12}$$

Note that the averages in this formula are not connected moments, which are distinguished by $\langle \rangle_c$.

Equation (12) yields only the continuum limit terms. To add Poisson noise contribution to the error, note that it arises from the possible overlaps between the indices (indices between two pairweights *can* still overlap!). In the spirit of infinitesimal Poisson models, we replace each overlap with a $1/\lambda$ term, and express the results in terms of connected moments. There are three possibilities: (i) no overlap (continuum limit)

$$f_{12}^a f_{34}^b (\xi_{1234} + \xi_{13}\xi_{24} + \xi_{14}\xi_{23}), \tag{13}$$

(ii) one overlap (4 possibilities)

$$\frac{4}{\lambda} f_{12}^a f_{13}^b (\xi_{123} + \xi_{23}), \tag{14}$$

(iii) two overlaps (2 possibilities)

$$\frac{2}{\lambda^2} f_{12}^a f_{12}^b (1 + \xi_{12}), \tag{15}$$

In these equations, for the sake of the Einstein convention we used $\xi(i,j,k,l) \rightarrow \xi_{ijkl}$. The above substitutions (rigorously true only in the infinitesimal Poisson sampling limit) become increasingly accurate with decreasing cell size. In that limit, adding the above three equations is equivalent to Eq. (8).

Dark Matter Caustics

P. SIKIVIE AND W. KINNEY

Department of Physics, University of Florida, Gainesville, Forida 32611, USA

ABSTRACT: The late infall of cold dark matter onto an isolated galaxy such as our own produces flows with definite velocity vectors at any physical point in the galactic halo. It also produces caustics which are places where the dark matter density is very large. The outer caustics are topological spheres whereas the inner caustics are rings. The self-similar model of galactic halo formation predicts that the caustic ring radii a_n follow the approximate law $a_n \sim 1/n$. In a recent study of 32 extended and well-measured galactic rotation curves, we found evidence for this law.

KEYWORDS: Caustics; Dark matter; Galactic halos; Rotation curves

1. INTRODUCTION

Before the onset of galaxy formation but after the time t_{eq} of equality between matter and radiation, the velocity dispersion of the cold dark matter candidates is very small, of order $\delta v_a(t) \sim 3 \cdot 10^{-17} \, (10^{-5} \, eV/m_a t) \, t(t_0/t)^{2/3}$ for axions and $\delta v_W(t) \sim 10^{-11} \, t(GeV/m_W)^{1/2} \, (t_0/t)^{2/3}$ for WIMPs, where t_0 is the present age of the universe and m_a and m_W are the masses of the axion and the WIMP, respectively. In the context of galaxy formation, such small velocity dispersions are entirely negligible. Massive neutrinos, on the other hand, have primordial velocity dispersion $\delta v_\nu(t) \simeq 5.3 \sim 10^{-4} \, eV/m_\nu) \sim (t_0/t)^{2/3}$ which is comparable to the virial velocity in galaxies and therefore nonnegligible in the context of galaxy formation [1], [2], [3]. This is the reason why massive neutrinos are called "hot dark matter."

Collisionless dark matter particles lie on a thin 3-dimensional (3D) sheet in 6D phase-space. The thickness of this sheet is the primordial velocity dispersion δv. If each of the aforementioned species of collisionless particles is present, the phase-space sheet has three layers, a very thin layer of axions, a medium layer of WIMPs and a thick layer of neutrinos. The phase-space sheet is located on the 3D hypersurface of points $(\vec{r}, \vec{v})$: $\vec{v} = H(t)\vec{r} + \Delta\vec{v} \, (\vec{r}, t)$ where $H(t) = 2/3t$ is the Hubble expansion rate and $\Delta\vec{v} \, (\vec{r}, t)$ is the peculiar velocity field. FIGURE 1 shows a 2D section of 6D phase-space along the $(z, \dot{z})$ plane. The wiggly line is the intersection of the 3D sheet on which the particles lie in phase-space with the plane of the figure. The thickness of the line is the velocity dispersion δv, whereas the amplitude of the wiggles in the line is the peculiar velocity Δv. If there were no peculiar velocities, the line would be straight since $\dot{z} = H(t)z$ in that case.

Address for correspondence: Department of Physics, New Physics Building, Corner N-S & Museum, University of Florida, Gainesville, Florida 32611, USA. Voice: 352/392-1923; fax: 352/392-8743.

sikivie@phys.ufl.edu, kinney@phys.ufl.edu

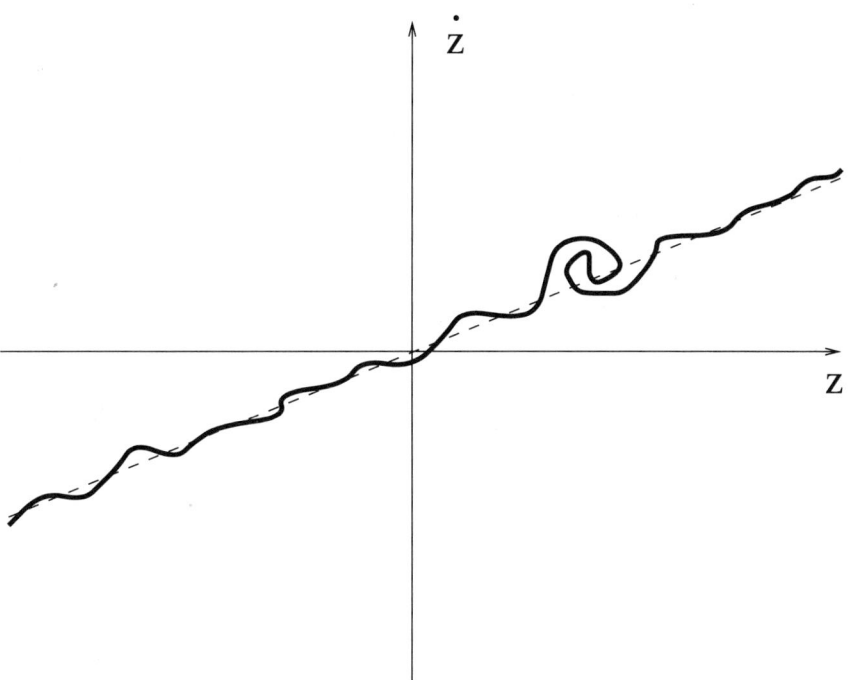

FIGURE 1. The wiggly line represents the intersection of the $(z, \dot{z})$ plane with the 3D sheet on which the collisionless dark matter particles lie in phase-space. The thickness of the line is the primordial velocity dispersion. The amplitude of the wiggles in the $\dot{z}$ direction is the velocity dispersion associated with density perturbations. Where an overdensity grows in the nonlinear regime, the line winds up in clockwise fashion. One such overdensity is shown.

The peculiar velocities are associated with density perturbations and grow by gravitational instability as $\Delta v \sim t^{2/3}$. On the other hand the primordial velocity dispersion decreases on average as $\delta v \sim t^{-2/3}$, consistently with Liouville's theorem. When an overdensity enters the nonlinear regime, the particles in its vicinity fall back onto it. This implies that the phase-space sheet 'winds up' there in clockwise fashion. One such overdensity is shown in FIGURE 1. In the linear regime, there is only one value of velocity, that is, one single flow, at a typical location in physical space, because the phase-space sheet covers physical space only once. On the other hand, inside an overdensity in the nonlinear regime, the phase-space sheet covers physical space multiple times implying that there are several (but always an odd number of) flows at such locations.

At the boundary surface between two regions one of which has n flows and the other $n + 2$ flows, the physical space density is very large because the phase-space sheet has a fold there. At the fold, the phase-space sheet is tangent to velocity space and hence, in the limit of zero velocity dispersion ($\delta v = 0$), the density diverges since it is the integral of the phase-space density over velocity space. The structure associated with such a phase-space fold is called a "caustic." It is a surface in physical space. It is easy to show that, in the limit of zero velocity dispersion, the density diverges as $d \sim 1/\sqrt{\sigma}$ when the caustic is approached from the side with $n + 2$ flows, where σ is the distance to the caustic. Velocity dispersion cuts off the divergence.

As mentioned above, the process of galactic halo formation involves the local winding up of the phase-space sheet of collisionless dark matter particles. If the galactic center is approached from an arbitrary direction at a given time, the local number of flows increases. First, there is one flow, then three flows, then five, seven, ... The number of flows at our location in the Milky Way galaxy today has been estimated [4] to be of order 100. The boundary between the region with one (three, five, ...) and the region with three (five, seven, ...) flows is the location of a caustic which is topologically a sphere surrounding the galaxy. When these caustic spheres are approached from the inside the density diverges as $d \sim 1/\sqrt{\sigma}$ in the zero velocity dispersion limit. These spheres are the outer caustics in the phase-space structure of galactic halos. In addition there are inner caustics.

It is a little more difficult to see why there must be inner caustics, and to derive their structure [4], [5]. The inner caustics are rings. They are located near where the particles with the most angular momentum in a given in-and-out flow are at their distance of closest approach to the galactic center. A ring is a closed tube whose cross-section is a D_{-4} catastophe [7]. The cross section is shown in FIGURE 2 in the limit

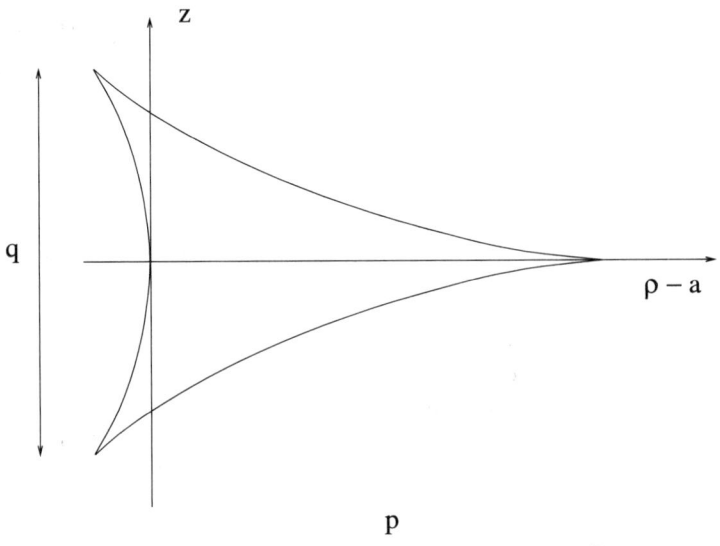

FIGURE 2. Cross section of a caustic ring in the case of axial and reflection symmetry. The galactic center is to the left of the figure. The z-direction is orthogonal to the galactic plane. The ρ-direction is radial. a is the caustic ring radius. The closed line with three cusps is the location of a caustic surface. The density diverges when the surface is approached from the inside as $\sigma^{-1/2}$ where σ is the distance to the surface.

of axial and reflection symmetry, and where the transverse dimensions, p and q, are much smaller than the ring radius a. In the absence of any symmetry, the cross-section of the tube does not have the exact shape shown in FIGURE 2 but it still has that shape qualitatively, that is, it is still a closed line with three cusps one of which points away from the galactic center.

The existence of caustic rings of dark matter follows from only two assumptions:

(1) the existence of collisionless dark matter
(2) that the velocity dispersion of the infalling dark matter is much less, by a factor ten say, than the rotation velocity of the galaxy.

Only the second assumption requires elaboration. Velocity dispersion has the effect of smoothing out caustics. The question is when is the velocity dispersion so large as to smooth caustic rings over distance scales of order the ring radius a, thus making the notion of caustic ring meaningless. This critical velocity dispersion was estimated [5] to be 30 km/s = 10^{-4} for the caustic rings in our own galaxy, whose rotation velocity is 220 km/s. 10^{-4} is much less than the *primordial* velocity dispersion δv of the cold dark matter candidates. However, the velocity dispersion Δv associated with density perturbations also smoothes caustics in coarse grained observations. So the question is whether the velocity dispersion Δv of cold dark matter particles associated with density perturbations falling onto our galaxy is less than 30 km/s. The answer is yes with high probability since the infalling dark matter particles are not associated with any observed inhomogeneities. 30 km/s is of order the velocity dispersion of the Magellanic Clouds. For the velocity dispersion of the dark matter particles presently falling onto our galaxy to be as large 30 km/s, these particles would have to be part of clumps whose mass/size ratio is of the order of the Magellanic Clouds. But if that were the case, why did these clumps fail to become luminous?

One might ask whether caustic rings can be seen in N-body simulations of galaxy formation. The generic surface caustics associated with simple folds of the phase-space sheet have been seen [8]–[11]. However, caustic rings would require far greater resolution than presently available, at least in a 3D simulation of our own halo. Indeed, the largest ring in our galaxy has been estimated [5] to have radius of order 40 kpc. It is part of an in-and-out flow that extends to the Galaxy's current turn-around radius, of order 2 Mpc. To resolve this first ring, the spatial resolution would have to be considerably smaller than 10 kpc. Hence, a minimum of $2 \cdot 1/(10 \text{ kpc})^3 (4\pi/3)(2\text{Mpc})^3 \simeq 7 \cdot 10^7$ particles would be required to see the caustic ring in a simulation of this one flow. However, the number of flows at 40 kpc in our halo [14], [15] is of order 10. So it appears that 10^9 particles is a strict minimum in a 3D simulation of our halo. Even so, this addresses only the kinematic requirement of resolving the halo in phase-space, assuming moreover that the particles are approximately uniformly distributed on the phase-space sheets. There is a further dynamical requirement that 2-body collisions do not artificially "fuzz up" the phase-space sheets. Indeed 2-body collisions are entirely negligible in the flow of cold dark matter particles such as axions or WIMPs. On the other hand, 2-body collisions are present and, hence, the velocity dispersion is artificially increased in the simulations. This may occur to such an extent that the caustics are washed away even if 10^9 particles are used.

In the self-similar infall model [12]–[15] of galactic halo formation the caustic ring radii a_n are predicted [5]:

$$\{a_n: n = 1, 2, \ldots\} \simeq (39, 19.5, 13, 10, 8, \ldots) \text{ kpc} \left(\frac{j_{\max}}{0.25}\right)\left(\frac{0.7}{h}\right)\left(\frac{v_{\text{rot}}}{220 \text{ km/s}}\right) \quad (1)$$

where h is the present Hubble rate in units of 100 km / sMpc, v_{rot} is the rotation velocity of the galaxy and $j_{\max}$ is the maximum of its dimensionless angular momentum distribution [14], [15]. In Eq. (1) we assume that the parameter [12], [15] $\varepsilon = 0.3$.

Equation (1) predicts the caustic ring radii of a galaxy in terms of its first ring radius a_1. If the caustic rings lie close to the galactic plane they cause bumps in the rotation curve, at the caustic ring radii. As a possible example of this effect, consider [5] the rotation curve of NGC3198, one of the best measured. It has three faint bumps at radii: 28, 13.5, and 9 kpc, assuming $h = 0.75$. The ratios happen to be consistent with Eq. (1) assuming the bumps are caused by the first three ($n = 1, 2, 3$) ring caustics of NGC3198. Moreover, since $v_{\text{rot}} = 150$ km/s, $j_{\max}$ is determined to equal 0.28. The uncertainty in h is a systematic effect that can be corrected for when determining $j_{\max}$ because the bump radii scale like $1/h'$ where h' is the Hubble rate assumed by the observer in constructing the rotation curve, and the caustic ring radii scale as $1/h$. Rises in the inner rotation curve of the Milky Way were also interpreted [5] as due to caustics $n = 6, 7, 8, 9, 10, 11, 12,$ and 13. This determined the value of $j_{\max}$ of our own galaxy to be 0.263. The first five caustic ring radii in our galaxy are then predicted to be: 41, 20, 13.3, 10, 8 kpc.

2. EVIDENCE FOR UNIVERSAL STRUCTURE IN GALACTIC HALOS

Motivated by these findings, we analyzed [16] a set of 32 extended well-measured galactic rotation curves which had been previously selected [17], [18] under the criteria that each is an accurate tracer of the galactic radial force law, and that it extends far beyond the edge of the luminous disk.

According to the self-similar caustic ring model, each galaxy has its own value of $j_{\max}$. Over the set of 32 galaxies selected [17], [18] $j_{\max}$ has some unknown distribution. However, the fact that the values of $j_{\max}$ of NGC3198 and of the Milky Way happen to be close to one another, within 7%, suggests that the $j_{\max}$ distribution may be peaked near a value of 0.27. Our strategy is to rescale each rotation curve according to

$$r \to \tilde{r} = r\left(\frac{220 \text{ km/s}}{v_{\text{rot}}}\right) \quad (2)$$

and to add them in some way. Because Eq. (1) predicts the nth caustic radius a_n to be distributed like $j_{\max}$ for all n, and it fixes the ratios $a_n/a_1 \simeq 1/n$, the sum of rotation curves should show the $j_{\max}$ distribution, once for $n = 1$, then at about half the $n = 1$ radii for $n = 2$, then at about 1/3 the $n = 1$ radii for $n = 3$, and so on. If the $j_{\max}$ dis-

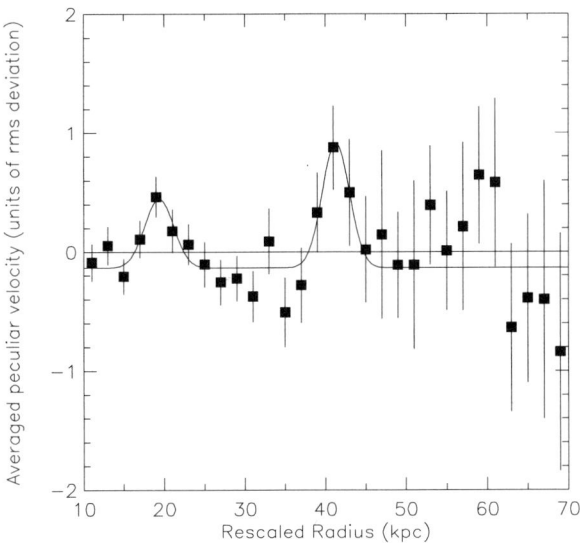

FIGURE 3. Binned data for 32 galaxy sample, with peaks fit to Gaussians.

tribution is broad, the sum of rotation curves is unlikely to show any feature. However, if it is peaked, then the sum should show a peak for $n = 1$ at some radius, then again at 1/2 that radius for $n = 2$, at 1/3 the radius for $n = 3$, and so on. If the j_{max} distribution is peaked at 0.263 (the value for the Milky Way) the peaks in the sum of rotation curves should appear at 41 kpc, 20 kpc, 13.3 kpc, ….

The procedure to add the 32 rotation curves [16] may be briefly described as follows. For each rotation curve, all data points with rescaled radii $\tilde{r} < 10$ kpc were deleted to remove the effect of the luminous disk. The remaining points were then fitted to a line. The rotation velocity v_{rot} used to rescale the radii in Eq. (2) is the average of that line. The rms deviation $\sqrt{\langle \delta v^2 \rangle}$ from the linear fit was determined for each galaxy. This was taken to be the error on the residuals δv_j, that is, the differences between the measured velocities in a rotation curve and the linear fit. Finally the sample of 32 galaxies was averaged in 2 kpc radial bins:

$$b_i \equiv \frac{1}{N_i} \sum_{j=1}^{N_i} \delta \tilde{v}_j \qquad (3)$$

where N_i is the number of data points in the bin. The assigned error on each b_i is then simply $1/\sqrt{N_i}$. FIGURE 3 shows the result.

There are two features evident at roughly 20 and 40 kpc. A fit to two Gaussians plus a constant indicates features at 19.4 ± 0.7 kpc and 41.3 ± 0.8 kpc, with overall significance of 2.4σ and 2.6σ, respectively. FIGURE 3 shows the fitted curve. When the same fit is applied to the same data in 1 kpc bins, the significance of the two peaks is 2.6 and 3.0 σ, respectively. The locations of the features agrees with the predictions of the self-similar caustic ring model with the j_{max} distribution peaked at 0.27. The use of Gaussians to fit the peaks in the combined rotation curve was an arbitrary choice in the absence of information on the j_{max} distribution.

The existence of velocity peaks and caustic rings in the cold dark matter distribution is relevant to axion [19] and WIMP searches [20]. Caustics may also be investigated using gravitational lensing techniques [21].

ACKNOWLEDGMENTS

This work was supported in part by the U.S. Department of Energy under grant No. DEFG05-86ER40272.

REFERENCES

1. TREMAINE, S. & J.E. GUNN. 1979. Dynamical role of light neutral leptons in cosmology. Phys. Rev. Lett. **42**: 407–4110.
2. BOND, J.R., G. EFSTATHIOU & J. SILK. 1980. Massive neutrinos and the large-scale structure of the universe. Phys. Rev. Lett. **45**: 1980–1984.
3. WHITE, S.D.M., C.S. FRENK & M. DAVIS. 1983. Clustering in a neutrino-dominated universe. Astrophys. J **274**: L1–L5.
4. SIKIVIE, P. & J.R. IPSER. 1992. Phase-space structure of cold dark matter halos. Phys. Lett. **B291**: 288–292.
5. SIKIVIE, P. 1998. Caustic rings of dark matter. Phys. Lett. B **432**: 139–144.
6. SIKIVIE, P. 1999. The caustic ring singularity. Phys. Rev. D **60**: 063501.
7. GILMORE, R. 1981. Catastrophy Theory for Scientists and Engineers. J. Wiley & Sons, Inc. New York.
8. DOROSHKEVICH, A.G. et al. 1980. Two-dimensional simulation of the gravitational system dynamics and formation of the large-scale structure of the universe. Mon. Not. R. Astr. Soc. **192**: 321–337.
9. KLYPIN, A.A. & S.F. SHANDARIN. 1983. Three-dimensional numerical model of the formation of large-scale structure in the universe. Mon. Not. R. Astr. Soc. **204**: 891–907.
10. CENTRELLA, J.M. & A.L. MELOTT. 1983. Three-dimensional simulation of large-scale structure in the universe. Nature **305**: 196–198.
11. MELOTT, A.L. & S.F. SHANDARIN. 1990. Generation of large-scale cosmological structures by gravitational clustering. Nature **346**: 633–635.
12. FILMORE, J.A. & P. GOLDREICH. 1984. Self-similar gravitational collapse in an expanding universe. Astrophys. J. **281**: 1–8.
13. BERTSCHINGER, E. 1985. Self-similar secondary infall and accretion in an Einstein-de Sitter universe. Ap. J. Suppl. **58**: 39–65.
14. SIKIVIE, P., I.I. TKACHEV & Y. WANG. 1995. The velocity peaks in the cold dark matter spectrum in Earth. Phys. Rev. Lett. **75**: 2911–2915.
15. SIKIVIE, P., I.I TKACHEV & Y. WANG. 1997. The secondary infall model of galactic halo formation and the spectrum of cold dark matter particles on Earth. Phys. Rev. D **56**: 1863–1878.
16. KINNEY, W. & P. SIKIVIE. 2000. Evidence for universal structure in galactic halos. Phys. Rev. **D61**: 087305.
17. BEGEMAN, K.G., A. H. BROEILS & R. H. SANDERS. 1991. Extended rotation curves of spiral galaxies — Dark haloes and modified dynamics. Mon. Not. R. Astr. Soc. **249**: 523-537.
18. SANDERS, R.H. 1996. The published extended rotation curves of spiral galaxies: confrontation with modified dynamics. Astrophys. J. **473**: 117–129.

19. SIKIVIE, P. 1999. Velocity peaks and caustic rings. *In* The Identification of Dark Matter. N. Spooner & V. Kudryavtsev, Eds.: 58–65. World Scientific Publishing Co. Singapore.
20. COPI, C., J. HEO & L. KRAUSS. 1999. Directional sensitivity, wimp detection, and the galactic halo. Phys. Lett. B **461:** 43–48.
21. HOGAN, C. 1999. Gravitational lensing by cold dark matter catastrophes. Astrophys. J. **527:** 42–45.

Nonlinear Gravitational Growth Inside and Outside the Standard Cosmology

E. GAZTAÑAGA AND J.A. LOBO

*INAOE, Coordinación de Astrofísica, Tonantzintla, Cholula,
Apdo Postal 216 y 51, 7200, Puebla, Mexico
Departament de Fisica Fonamental, Universitat de Barcelona,
Diagonal 647, 08028 Barcelona, Spain*

ABSTRACT: We reconsider the problem of nonlinear structure formation inside and outside General Relativity (GR), both in the weakly and strongly nonlinear regime. We show how these regimes can be explored observationally through clustering of high-order cumulants and through the epoch of formation, abundance and clustering of collapse structures, using Press and Schechter (1974, Astrophys. J. 187: 425–438) formalism and its extensions.

KEYWORDS: Galaxies: formation, gravitation, instabilities; Large-scale structure of the Universe

1. INTRODUCTION

In cosmology, the standard picture of gravitational growth and also many aspects of fundamental physics are extrapolated many orders of magnitude, from the scales and times where our current theory of gravity (general relativity, GR) has been experimentally tested, into the distant universe. It is important to evaluate how much our predictions and cosmological picture depend on the underlying hypothesis.

One aspect of GR that could be questioned or tested without modifying the basic structure or symmetry of the theory is Einstein's field equations, relating the energy content ($T_{\mu\nu}$) to the curvature ($R_{\mu\nu}$). One such modification, which will be considered here, is scalar-tensor theories (STT), such as Brans-Dicke (BD) theory. A more generic, but also more vague, way of testing the importance of Einstein's field equations is to model independently the geometry and the matter content, thus allowing for the possibility of other relations between them.

To address the above question we will consider two nonstandard variations: scalar-tensor models and some examples of a cosmology that do not obey Einstein's field equations. We consider three main regimes for structure formation in nonstandard gravity: linear, weakly nonlinear, and strongly nonlinear large scale clustering. We restrict to the shear-free or spherical collapse (SC) model, which corresponds to the spherically symmetric (or local) dynamics (see below). This ap-

---

Address for correspondence: Enrique Gaztañaga, INAOE, Coordinación de Astrofiscia, Tonantzintla, Cholula, Apdo Postal 216 y 51, 7200, Puebla, Mexico. Voice: +52 22 47 20 11 x.2303; fax: +52 22 47 22 31.
gazta@inaoep.mx

proximation works well at least in the two different observational contexts, that will be explored here: higher-order cumulants in the galaxy distribution and the epoch of formation and abundance of structures using the Press and Schechter [1] formalism.

In Section 2 we introduce the notation and give a brief summary of how structure formation relates to the underlying theory of Gravity. We also present the more general case of an ideal (relativistic) fluid. In Section 3 we show how these predictions change in the two examples of nonstandard variations on GR. Observational consequences and the conclusions are presented in Section 4.

2. GRAVITATIONAL GROWTH INSIDE GR

The self-gravity of a local overdense region works against the expansion of the universe so that this region will expand at a slower rate that the background. This increases the density contrast so that eventually the region collapses. The details of this collapse depend on the initial density profile. Here we will focus in the spherically symmetric case. We will revise nonlinear structure growth in the context of the fluid limit and the shear-free or spherical collapse approximation. This turns out to be very good approximation for the applications that will be considered later (leading order and strongly nonlinear statistics).

2.1. Einstein's and Raychaudhuri's Equations

We start recalling that the metric tensor $g_{\mu\nu}$ defines the line element of spacetime: $ds^2 = g_{\mu\nu}dx^\mu dx^\nu$ which in the homogeneous and isotropic model of the cosmological principle can be written as:

$$ds^2 = dt^2 - a(t)^2\left[\frac{dr^2}{1+k^2r^2} + r^2(d\theta^2 + \sin^2\theta d\phi^2)\right]. \tag{1}$$

As usual we will work in comoving coordinates x related to physical coordinates by $r_p = a(t)x$, where $a(t) = (1 + z)^{-1}$ is the cosmic scale factor, and z the corresponding redshift ($a_0 \equiv 1$). Thus all geometrical aspects of this universal line element are determined up to the function $a(t)$ and the arbitrary constant k, which defines the usual open, flat, and closed universes. The function $a(t)$ can be found for each energy content by solving the corresponding equations of motion, for example, the gravitational field equations.

In this section we consider Einstein's equations:

$$R_{\mu\nu} + \Lambda g_{\mu\nu} = -8\pi G(T_{\mu\nu} - \frac{1}{2}g_{\mu\nu}T) \tag{2}$$

where $T \equiv g^{\mu\nu}T_{\mu\nu}$ is the trace of the energy-momentum tensor; we have included a cosmological constant term to keep the equations general at this stage. For an ideal fluid, we have:

$$T_{\mu\nu} = p g_{\mu\nu} + (p + \rho)u_\mu u_\nu \tag{3}$$

If these equations are expanded one finds that

$$\frac{3\ddot{a}}{a} = -4\pi G p\left(1 + \frac{3p}{\rho}\right) + \Lambda \tag{4}$$

$$\frac{\dot{a}^2}{a^2} = \frac{8\pi G p}{3} + \frac{k}{a^2} + \frac{\Lambda}{3}, \quad \dot{} \equiv \frac{d}{dt}. \tag{5}$$

In the fluid approximation, deviations from the mean background are characterized by fluctuations in the density and velocity fields. The continuity equation for a nonrelativistic fluid is [2]:

$$\frac{\partial}{\partial \tau}\delta(x,\tau) + \nabla \cdot \{[1 + \delta(x,\tau)]v(x,\tau)\} = 0 \tag{6}$$

where $\rho(x,\tau) = \delta(x,\tau)/\overline{\rho} - 1$ is the local *density contrast*, $v(x,\tau)$ the *peculiar velocity*, and τ the *conformal time* defined by

$$d\tau = \frac{dt}{a(t)} \Leftrightarrow \frac{d}{dt} = \frac{1}{a}\frac{d}{d\tau}. \tag{7}$$

The continuity equation (6) can also be written

$$\frac{d\delta}{d\tau} + (1 + \delta)\theta = 0, \quad \theta = \nabla \cdot v. \tag{8}$$

In order to find an equation of motion for the density contrast alone we shall resort to the Raychaudhuri equation (see, e.g., [3])

$$\frac{d\Theta}{ds} + \frac{1}{3}\Theta^2 = -\sigma_{ij}\sigma^{ij} + \omega_{ij}\omega^{ij} + R_{\mu\nu}u^\mu u^\nu \tag{9}$$

where $\Theta \equiv \nabla_\mu u^\mu$, σ_{ij} is the *shear* tensor, ω_{ij} the *vorticity* tensor, and $R_{\mu\nu}$ the Ricci tensor; u^μ is the fluid's 4-velocity, $u^0 = 1$, and

$$u = \dot{a}(t)x + v(x,t) \tag{10}$$

It is important to stress that Raychaudhuri's equation, Eq. (9), is *purely geometric*: it describes the evolution in proper time of the dilatation coefficient Θ of a bundle of nearby geodesics. There is no physics in this equation until a relationship between $R_{\mu\nu}$ and the matter contents of the universe is specified by means of a set of field equations. This makes it very useful for our purposes in this paper, as we shall later make reference to a different set of field equations.

If Einstein's field equations, Eqs. (2) and (3), are assumed then it is readily verified that

$$R_{\mu\nu}u^\mu u^\nu = -4\pi G\rho\left(1 + \frac{3p}{\rho}\right) + \Lambda \tag{11}$$

2.2. Shear Free and Matter Domination

In a matter dominated regime ($p = 0$), $\rho \sim a^{-3}$ and Eq. (5) can be rewritten using the the notation: $\Omega_M \equiv 8\pi G \rho_0/(3H_0^2)$ is the ratio of matter density to critical density, $\Omega_k = K/H_0^2$ gives the global curvature, and $\Omega_\Lambda = \Lambda/(3H_0^2)$ where Λ is the cosmological constant, so that $\Omega_M + \Omega_k + \Omega_\Lambda = 1$:

$$H^2(z) \equiv \left(\frac{\dot{a}}{a}\right)^2 = H_0^2[\Omega_M(1+z)^3 + \Omega_k(1+z)^2 + \Omega_\Lambda]. \tag{12}$$

We can now substitute Eq. (11) into Eq. (9). In a matter-dominated regime, and for a shear free, nonrotating cosmic fluid we obtain:

$$\frac{d\Theta}{d\tau} + \frac{1}{3}\Theta^2 = -4\pi G \rho + \Lambda. \tag{13}$$

On making use of Eq. (10) we can split Θ as

$$\Theta = \nabla_\mu u^\mu = \frac{3\ddot{a}}{a} + \frac{\theta}{a}, \tag{14}$$

so that, taking into consideration the field equations for the expansion factor $a(t)$ (Eqs. (4) and (5)), Eq. (13) can be recast in the form

$$\frac{d\theta}{d\tau} + \mathcal{H}(\tau) + \frac{1}{3}\theta^2 = -4\pi G a^2 \bar{\rho} \delta \tag{15}$$

where $\mathcal{H}(\tau) \equiv d(\ln a)/d\tau$. We can now eliminate θ between Eqs.(8) and (15) to find the following second-order differential equation for the density contrast:

$$\frac{d^2\delta}{d\tau^2} + \mathcal{H}(\tau)\frac{d\delta}{d\tau} - \frac{3}{2}\mathcal{H}^2(\tau)\Omega_M(\tau)\delta = \frac{4}{3}(1+\delta)^{-1}\left(\frac{d\delta}{d\tau}\right)^2 + \frac{3}{2}\mathcal{H}^2(\tau)\Omega_M(\tau)\delta^2, \tag{16}$$

where we have shifted to the rhs all nonlinear terms, and used the notation

$$\Omega_M(\tau) = \frac{\Omega_M}{\Omega_M + a\Omega_k + a^3\Omega_\Lambda}. \tag{17}$$

Equation (16) reproduces the the equation of the *spherical collapse* model (SC). In other words, *the SC approximation is the actual dynamics when shear is neglected*. As one would expect, this yields a *local* evolution, in the restricted sense that the evolved field at a point is just given by a local (nonlinear) transformation of the initial field at the same point.

2.3. Linear Growth

We next do a perturbative expansion for δ. The first contribution is the linear theory solution. For this, Eq. (16) clearly simplifies to

$$\frac{d^2\delta_l}{d\tau^2} + \mathcal{H}(\tau)\frac{d\delta_l}{d\tau} - \frac{3}{2}\mathcal{H}^2(\tau)\Omega_M(\tau)\delta_l = 0, \qquad (18)$$

where δ_l stands for the "linear" solution. Because this is a linear equation, the spatial and temporal part of $\delta = \delta(x,\tau)$ factorize:

$$\delta_1(x,\tau) = \delta_0(x)D(\tau) \qquad (19)$$

where D is usually called the *linear growth factor*. This means that initial fluctuations, no matter of what amplitude, grow by the same factor, and the statistical properties of the initial field are just linearly scaled. For example, the N-point correlation functions are:

$$\xi_N(r_1, ..., r_N, t) = D^N \xi_N(r_1, ..., r_N, 0) \qquad (20)$$

To find the solution to Eq. (18) it is expedient to change the time variable to $\eta = \ln(a)$, so that

$$\frac{d}{d\eta} = \frac{1}{\mathcal{H}(\tau)}\frac{d}{d\tau} = \frac{1}{H}\frac{d}{dt} \qquad (21)$$

We then have

$$\frac{d^2D}{d^2\eta} + \left(2 + \frac{\dot{H}}{H^2}\right)\frac{dD}{d\eta} - \frac{3}{2}\Omega_M(\eta)D = 0 \qquad (22)$$

where we can write

$$\frac{\dot{H}}{H^2} = -\frac{3}{2}\left(\frac{\Omega_M + 2/3\, e^\eta \Omega_k}{\Omega_M + e^\eta \Omega_k + e^{3\eta}\Omega_\Lambda}\right) \qquad (23)$$

$$\Omega_M(\eta) = \frac{\Omega_M}{\Omega_M + e^\eta \Omega_k + e^{3\eta}\Omega_\Lambda}, \qquad (24)$$

where Ω_M, Ω_k, and Ω_Λ are just constants (the current value at $a = 1$).

In the Einstein–deSitter universe ($\Omega_k = \Omega_\Lambda = 0$) we have that $\Omega_M(\eta) = 1$ and $\dot{H}/H^2 = -3/2$, so the differential equation becomes

$$\frac{d^2D}{d^2\eta} + \frac{1}{2}\frac{dD}{d\eta} - \frac{3}{2}D = 0 \qquad (25)$$

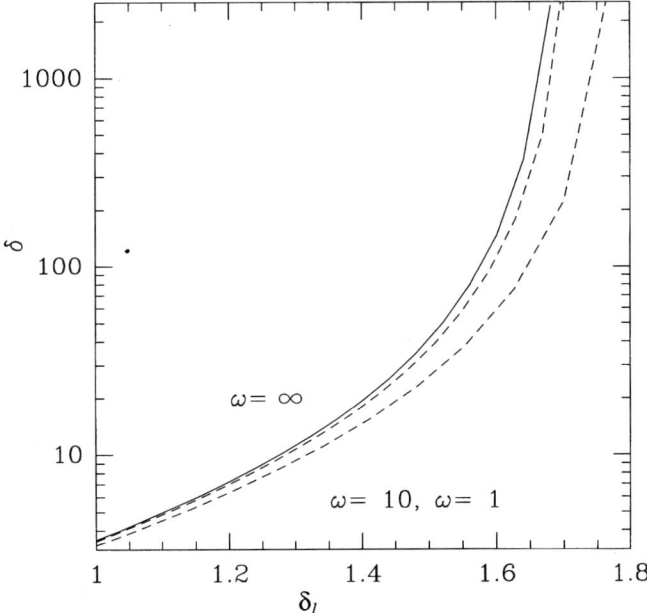

FIGURE 1. The nonlinear density contrast, δ, as a function of the linear one δ_l in the spherical collapse. The *continuous line* shows the GR prediction ($\omega = \infty$), the *short-dashed lines* correspond to the same solution in the BD model with $\omega = 10$ and $\omega = 1$ (from left to right).

whose solutions

$$D = C_1 e^{\eta} + C_2 e^{-3/2\eta} = C_1 a + C_2 a^{-3/2} \tag{26}$$

reproduce the usual linear growth $D \sim a$ and the decaying solutions $D \sim a^{-3/2}$.

2.4. Nonlinear Growth

The exact (nonperturbative) solution for the SC Eq. (16) for the density contrast in an Einstein–deSitter universe admits a well-known exact solution (see [2]). There is also a solution for the $\Omega_M \neq 1$ case (see [4], [5]). The continuous line in FIGURE 1 illustrates the solution to the above equation (the other lines will be explained later). Note the singularity at $\delta_c \simeq 1.68$, which corresponds to the gravitational collapse (see [2]). In the Press and Schechter [1] formalism and its extensions the value of δ_c marks to the value of the linear overdensity at the time of collapse. With these models and the value of δ_c we can predict the evolution of the mass function of halos and also their clustering properties.

If we are only interested in the perturbative regime ($\delta_1 \to 0$), which is the relevant one for the description of structure formation on large scales, the above solution can also be expressed directly in terms of the initial density contrast, δ_0, which plays the role of the initial size of the spherical fluctuation in Eq. (19). This way, the evolved density contrast in the perturbative regime is given by a *local-density* transformation of the linear density fluctuation,

$$\delta = f(\delta_1) = \sum_{n=1}^{\infty} \frac{v_n}{n!}[\delta_1]^n \qquad (27)$$

Notice that all the dynamical information in the SC model is encoded in the v_n coefficients of this *local-density transformation*, Eq. (27).

We can introduce the latter power series expansion in Eq. (16) and determine the v_n coefficients one by one. Before we do this, it is convenient to change again the time variable to $\eta = \ln(a)$ as we did in the linear case, Eq. (22):

$$\frac{d^2\delta}{d^2\eta} + \left(2 + \frac{\dot{H}}{H^2}\right)\frac{d\delta}{d\eta} - \frac{3}{2}\Omega_M(\eta)\delta = \frac{4}{3}\frac{1}{1+\delta}\left(\frac{d\delta}{d\eta}\right)^2 + \frac{3}{2}\Omega_M(\eta)\delta^2 \qquad (28)$$

We can now use the expansion in Eq. (27) with δ_1 given by the linear growth factor $D = a = e^\eta$ and compare order by order. For the Einstein–deSitter universe they turn out to be

$$v_2 = \frac{34}{21} \sim 1.62$$

$$v_3 = \frac{682}{189} \sim 3.61$$

$$v_4 = \frac{446440}{43659} \sim 10.22 \qquad (29)$$

$$v_5 = \frac{8546480}{243243} \sim 35.13$$

and so on (see, e.g., [5] for other cases). Once we have these coefficients we can get the evolution of the nonlinear variance and higher order moments in terms of the initial conditions for either Gaussian (e.g., [4], [6]) or non-gaussian [7] initial conditions.

2.5. *Equation of State* $p = \gamma\rho$

The time-component of the energy conservation equation $\nabla_\nu T^{\mu\nu} = 0$ gives us (for $p = \gamma\rho$) both the background density behavior

$$\bar{\rho}a^{3(1+\gamma)} = \text{const} \qquad (30)$$

and the continuity equation for the density contrast

$$\frac{d\delta}{d\tau} + (1+\gamma)(1+\delta)\theta = 0 \qquad (31)$$

where, like before, τ is the conformal time, and $\theta \equiv \nabla \cdot \mathbf{v}$. This is the generalization of Eq. (8) for a relativistic fluid. Also, Hubble's equation, Eq. (12), now becomes

$$H^2 = H_0^2[\Omega_M a^{-3(1+\gamma)} + \Omega_k a^{-2} + \Omega_\Lambda] \qquad (32)$$

We can combine Eq. (31) with the Raychaudhuri equation for this case (cf. Eqs. (9) and (11))

$$\frac{d\Theta}{dt} + \frac{1}{3}\Theta^2 = -4\pi G\rho(1 + 3\gamma) + \Lambda \tag{33}$$

to obtain, after some algebra,

$$\frac{d^2\delta}{d^2\eta} + \left(2 + \frac{\dot{H}}{H^2}\right)\frac{d\delta}{d\eta} - \frac{3}{2}(1 + \gamma)(1 + 3\gamma)\Omega(\eta)\delta =$$
$$\frac{4 + 3\gamma}{3 + 3\gamma}\frac{1}{1 + \delta}\left(\frac{d\delta}{d\eta}\right)^2 + \frac{3}{2}(1 + \gamma)(1 + 3\gamma)\Omega(\eta)\delta^2 \tag{34}$$

where we have expediently redefined $\Omega(\eta)$ in Eq. (24) to

$$\Omega_M(\eta) = \frac{\Omega_M}{\Omega_M + e^{\eta(1 + 3\gamma)}\Omega_k + e^{3\eta(1 + \gamma)}\Omega_\Lambda}, \tag{35}$$

and we can write

$$\frac{\dot{H}}{H^2} = -\frac{3}{2}\left(\frac{(1 + \gamma)\Omega_M e^{-3\eta\gamma} + 2/3\, e^\eta\Omega_k}{\Omega_M e^{-3\eta\gamma} + e^\eta\Omega_k + e^{3\eta}\Omega_\Lambda}\right). \tag{36}$$

In an Einstein–deSitter universe ($\Omega_k = \Omega_\Lambda = 0$), $\Omega(\eta) = 1$, and the linear regime is governed by

$$\frac{d^2D}{d^2\eta} + \frac{1 - 3\gamma}{2}\frac{dD}{d\eta} - \frac{3}{2}(1 + \gamma)(1 + 3\gamma)D = 0 \tag{37}$$

which has the usual solutions of the form $D = a^\alpha$, with

$$\alpha_1 = 1 + 3\gamma, \quad \alpha_2 = -3(1 + \gamma)/2 \tag{38}$$

FIGURE 2 shows these perturbative solutions. The growing mode for $\gamma > -1/3$ is:

$$\alpha_1 = 1 + 3\gamma \tag{39}$$

$$v_2 = \frac{2(17 + 48\gamma + 27\gamma^2)}{3(1 + \gamma)(7 + 15\gamma)} \tag{40}$$

$$v_3 = \left[72 + 540\gamma + 324\gamma^2 + \frac{16}{(1 + \gamma)^2} + \frac{24}{1 + \gamma} - \frac{(6 + 18\gamma)(17 + 48\gamma + 27\gamma^2)}{(1 + \gamma)(7 + 15\gamma)}\right](27 + 144\gamma + 189\gamma^2)^{-1} \tag{41}$$

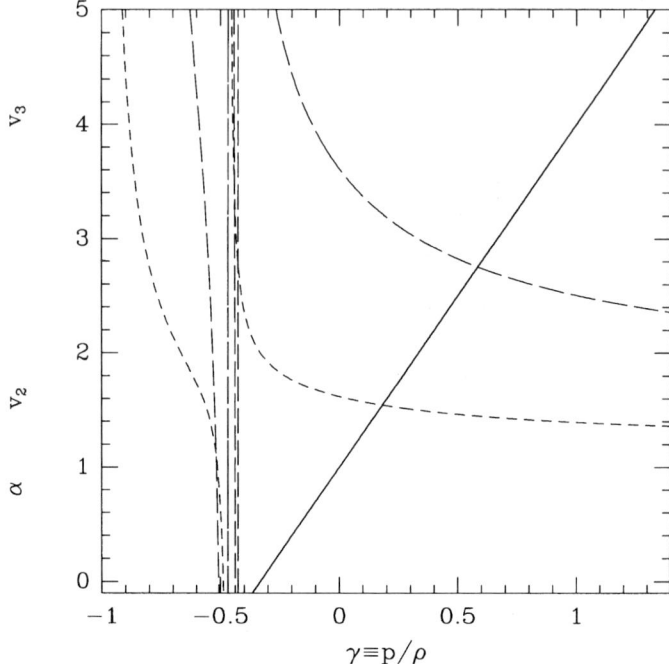

FIGURE 2. The linear growth index α_1 (*continuous line*) and nonlinear coefficients v_2 (*short-dashed*) and v_3 (*long-dashed*), as a function of the equation of state $\gamma \equiv p/\rho$. As can be seen from Eq. (39)–(44), v_2 and v_3 have a very rapid variation (discontinuities) around $\gamma \simeq -0.5$. There is no growing modes for $-1 < \gamma < -1/3$.

while for $\gamma < -1$ (negative pressure):

$$\alpha_2 = \frac{-3(1+\gamma)}{2} \tag{42}$$

$$v_2 = \frac{3}{2} \tag{43}$$

$$v_3 = 3 \tag{44}$$

For radiation ($\gamma = 1/3$) we have that $\alpha_1 = 2$, which reproduces the well-known results (see [2]) and $v_2 = 3/2$ and $v_3 = 3$, which are new results as far as we know. Note that these values are identical to the case of negative pressure, $\gamma < -1$, the only difference being in the linear growth, but for $\gamma = -7/3$ all α, v_2, and v_3 are identical to the radiation case. In the limit of strong pressure $\gamma \to \infty$ we find: $v_2 = 6/5$ and $v_3 = 12/7$. As can be seen in FIGURE 2, and also in the equations above, there are poles for v_2 at $\gamma = -1$ and $\gamma = -7/15$.

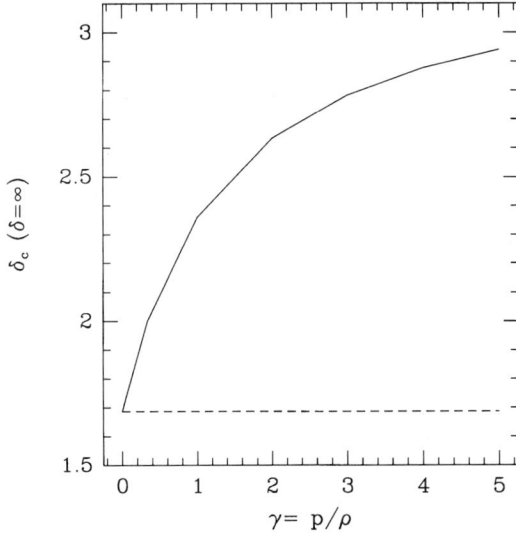

FIGURE 3. The critical value of the linear density contrast δ_c where $\delta = \infty$ as a function of $\gamma \equiv p/\rho$.

FIGURE 3 shows the variation in δ_c, defined as the value of the linear overdensity where the corresponding nonlinear value becomes infinity.

3. GRAVITATIONAL GROWTH OUTSIDE GR: SCALAR-TENSOR THEORIES

Here we investigate how a varying G could change the above results. We parametrize the variation of G using scalar-tensor theories (STT) of gravity such as Brans–Dicke (BD) theory or its extensions.

To make quantitative predictions we will consider cosmic evolution in STTs, where G is derived from a scalar field ϕ which is characterized by a function $\omega = \omega(\phi)$ determining the strength of the coupling between the scalar field and gravity. In the simplest BD models, ω is just a constant and $G \simeq \phi^{-1}$ (see below). However, if ω varies then it can change with cosmic time, so that $\omega = \omega(z)$. The structure of the solutions to BD equations is quite rich and depends crucially on the coupling function $\omega(\phi)$ (see [8]).

Here we shall be considering the standard BD model with constant ω; the field equations are (see, e.g., [9]):

$$R_{\mu\nu} = -\frac{8\pi}{\phi}\left(T_{\mu\nu} - \frac{1+\omega}{3+2\omega}g_{\mu\nu}T\right) - \frac{\omega}{\phi^2}\nabla_\mu\phi\nabla_\nu\phi - \frac{1}{\phi}\nabla_\mu\nabla_\nu\phi \quad (45)$$

$$\Box\phi = -\frac{8\pi}{3+2\omega}T, \quad (T \equiv g^{\mu\nu}T_{\mu\nu}). \tag{46}$$

The Hubble rate H for a homogeneous and isotropic background universe can be easily obtained from the above; it is:

$$H^2 \equiv \left(\frac{\dot{a}}{a}\right)^2 = \frac{8\pi\rho}{3\phi} + \frac{k}{a^2} + \frac{\Lambda}{3} + \frac{\omega\dot{\phi}^2}{6\phi^2} - H\frac{\dot{\phi}}{\phi}. \tag{47}$$

These equations must be complemented with the equation of state for cosmic fluid. In a flat, matter dominated universe ($p = 0$), an exact solution to the problem can be found:

$$G = \frac{4+2\omega}{3+2\omega}\phi^{-1} = G_0(1+z)^{-1/(1+\omega)} \tag{48}$$

and

$$a(t) = (t/t_0)^{(2\omega+2)/(3\omega+4)}. \tag{49}$$

This solution for the flat universe is recovered in a general case in the limit $t \to \infty$, and also arises as an exact solution of Newtonian gravity with a power law $G \propto t^n$ [10]. For nonflat models, $a(t)$ is not a simple power law and the solutions get far more complicated. To illustrate the effects of a nonflat cosmology we will consider general solutions that can be parametrized as Eq. (48) but which are not simple power-laws in $a(t)$. In this case, it is easy to check that the new Hubble law given by Eq. (47) becomes

$$H^2(z) = H_0^2 [\hat{\Omega}_M(1+z)^{3+1/(1+\omega)} + \hat{\Omega}_k(1+z)^2 + \hat{\Omega}_\Lambda)], \tag{50}$$

where $\hat{\Omega}_M$, $\hat{\Omega}_k$ and $\hat{\Omega}_\Lambda$ follow the usual relation $\hat{\Omega}_M + \hat{\Omega}_k + \hat{\Omega}_\Lambda = 1$, and are related to the familiar local ratios ($z \to 0$: $\Omega_M \equiv 8\pi G_0 \rho_0/(3H_0^2)$, $\Omega_k = k/H_0^2$ and $\Omega_\Lambda = \Lambda/(3H_0^2)$) by

$$\hat{\Omega}_M = \Omega_M \frac{3(1+\omega)^2}{(2+\omega)(4+3\omega)},$$

$$\hat{\Omega}_\Lambda = \Omega_\Lambda \frac{6(1+\omega)^2}{(3+2\omega)(4+3\omega)}, \tag{51}$$

$$\hat{\Omega}_k = \Omega_k \frac{6(1+\omega)^2}{(3+2\omega)(4+3\omega)}.$$

Thus the GR limit is recovered as $\omega \to \infty$.

We now investigate the density fluctuations in the above theory. As in Section 2, we shall make use of the continuity equation (8) in combination with the Raychaudhuri equation (9). As mentioned above (see Section 2.1), both of these are still valid within the context of BD theory: it is only needed to replace the Ricci tensor in the rhs of Eq. (9) according to BD's field equations, Eq. (45). Considering again a nonrotating, shear-free cosmic fluid, the following obtains:

$$\frac{d\Theta}{dt} + \frac{1}{3}\Theta^2 = -\frac{4+2\omega}{3+2\omega}\frac{4\pi\rho}{\phi}\left(1 + \frac{1+\omega}{2+\omega}\frac{3p}{\rho}\right) - \omega\frac{\dot\phi^2}{\phi^2} - \frac{\ddot\phi}{\phi}. \tag{52}$$

We shall still make use of a gravitational "constant" parametrized as in equation (48) above; this is justified insofar as the characteristic length for the variation of ϕ is typically much greater than that of the density fluctuations in a matter dominated universe (see, e.g., (Nariai 1969). In this approximation, Eq. (9) gives

$$\frac{d\theta}{dt} + \mathcal{H}(\tau)\theta + \frac{1}{3}\theta^2 = -\frac{4+2\omega}{3+2\omega}\frac{4\pi a^2\bar\rho\delta}{\phi}. \tag{53}$$

where τ is again the *conformal time* parameter, $d\tau = a^{-1}dt$, and θ is defined in Eqs. (8) and (14). Remarkably, this equation is very similar to the GR equation (3): it is only needed to replace in it the gravitational constant G by its expression as a multiple of the varying scalar field ϕ as given in Eq. (48). Combining (53) with the continuity equation (8) we immediately find

$$\frac{d^2\delta}{d\tau^2} + \mathcal{H}(\tau)\frac{d\delta}{d\tau} + \frac{4}{3}(1+\delta)^{-1}\left(\frac{d\delta}{d\tau}\right)^2 = \frac{4+2\omega}{3+2\omega}\frac{4\pi a^2\rho\delta}{\phi}. \tag{54}$$

As in Section 2, we change the independent variable in Eq. (54) to $\eta = \ln a$, whereby we obtain

$$\frac{d^2\delta}{d\eta^2} + \left(2 + \frac{\dot H}{H^2}\right)\frac{d\delta}{d\eta} - \frac{4}{3}(1+\delta)^{-1}\left(\frac{d\delta}{d\eta}\right)^2 = \frac{4+2\omega}{3+2\omega}\frac{4\pi a^2\rho\delta}{H^2\phi}. \tag{55}$$

Using Eq. (47) to calculate $\dot H$, and assuming further that $\hat\Omega_k = \hat\Omega_\Lambda = 0$, we finally get

$$\frac{d^2\delta}{d\eta^2} + \frac{1}{2} + \frac{1}{1+\omega}\frac{d\delta}{d\eta} - \frac{1}{2}\frac{(2+\omega)(4+3\omega)}{(1+\omega)^2}\delta = \frac{4}{3}(1+\delta)^{-1}\left(\frac{d\delta}{d\eta}\right)^2. \tag{56}$$

We next examine the solutions to this equation.

3.1. Linear Growth

Let us call $D(\eta)$ the solution to the linearized version of Eq. (56), that is,

$$\frac{d^2D}{d^2\eta} + \frac{1}{2}\frac{\omega}{1+\omega}\frac{dD}{d\eta} - \frac{1}{2}\frac{(2+\omega)(4+3\omega)}{(1+\omega)^2}D = 0 \tag{57}$$

Again the solutions are given by the roots α_1 and α_1 of the corresponding characteristic functions:

$$D = C_1 a^{\alpha_1} + C_2 a^{\alpha_2} \tag{58}$$

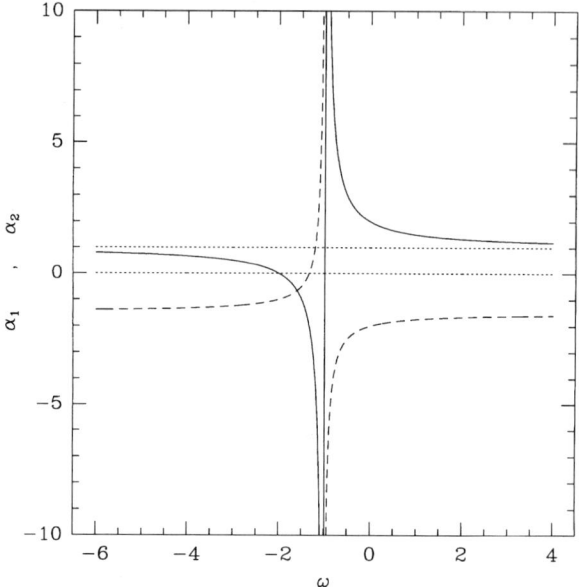

FIGURE 4. The linear growth indices α_1 (*continuous line*) and α_2 (*dashed line*) as a function of the BD parameter ω.

with

$$\alpha_1 = \frac{2+\omega}{1+\omega} \simeq 1 + \frac{1}{\omega} + \mathcal{O}\left(\frac{1}{\omega^2}\right) \tag{59}$$

$$\alpha_2 = \frac{-4-3\omega}{2+2\omega} \simeq -\frac{3}{2} - \frac{1}{2}\frac{1}{\omega} + \mathcal{O}\left(\frac{1}{\omega^2}\right) \tag{60}$$

which reproduces the usual linear growth $D \sim a$ and $D \sim a^{-3/2}$ in the limit $\omega \to \infty$. Note that α_1 corresponds to the growing mode only for large values of $|\omega|$, but the situation is more complicated when ω is not large.

FIGURE 4 shows the values of α_1 and α_2 as functions of ω. The effective G in BD decreases as the Universe expands if $-1 < \omega < \infty$, and the expansion factor $a(t)$ stops for $\omega = -1$; the growing mode in this regime is controlled by α_2, since this is the positive root. The universe shrinks to an eventual collapse if $-4/3 < \omega < -1$ (see Eq. 49), in which case there are no growing modes, as can be seen in FIGURE 4 (both α_1 and α_2 are negative). For $\omega < -4/3$ the expansion factor grows with time again and α_2 becomes the growing mode again. Notice that in this regime of $\omega < -4/3$, $\alpha_2 < 1$, so that it is slower than for $\omega > 0$. As we will show this is compensated in part by a stronger nonlinear growth.

3.2. Nonlinear Growth

In the nonlinear case Eq. (28) for Einstein-deSitter turns into:

$$\frac{d^2\delta}{d^2\eta} + \frac{1}{2}\frac{\omega}{1+\omega}\frac{d\delta}{d\eta} - \frac{1}{2}\frac{(2+\omega)(4+3\omega)}{(1+\omega)^2}\delta = \frac{4}{3}\frac{1}{1+\delta}\left(\frac{d\delta}{d\eta}\right)^2 + \frac{1}{2}\frac{(2+\omega)(4+3\omega)}{1+\omega^2}\delta^2 \quad (61)$$

We can now proceed as before, using the expansion in Eq. (27) with δ_1 given by the linear growth factor $D = a^{\alpha_1} = e^{\alpha_1\eta}$, and compare order by order. We find:

$$\nu_2 = \frac{34\omega + 56}{21\omega + 36} = \frac{34}{21}\left[1 - \frac{8}{119}\frac{1}{\omega} + \mathcal{O}\left(\frac{1}{\omega^2}\right)\right] \quad (62)$$

$$\begin{aligned}\nu_3 &= \frac{2(944 + 1136\omega + 341\omega^2)}{3(12+7\omega)(16+9\omega)} \\ &= \frac{682}{189}\left[1 + \frac{3452}{21483}\frac{1}{\omega} + \mathcal{O}\left(\frac{1}{\omega^2}\right)\right]\end{aligned} \quad (63)$$

$$\begin{aligned}\nu_4 = &\left(1457344 + 1621552\omega - \frac{76672(28+17\omega)}{12+7\omega}\right.\\ &- \frac{38336\omega(28+17\omega)}{12+7\omega} + \frac{13440(28+17\omega)^2}{(12+7\omega)^2} \\ &+ 44644\omega^2 + \frac{6720\omega(28+17\omega)^2}{(12+7\omega)^2} \\ &\left.+ \frac{6272(944+1136\omega+341\omega^2)}{(12+7\omega)(16+9\omega)} + \frac{3136\omega(944+1136\omega+341\omega^2)}{(12+7\omega)(16+9\omega)}\right)\\ &\times (158760 + 166698\omega + 43659\omega^2)^{-1}\end{aligned} \quad (64)$$

Note how for positive ω nonlinear effects tend to compensate the increase in linear effects (see FIG. 4), whereas for $\omega < -4/3$, the linear effects are reduced ($\alpha < 1$) while nonlinearities get larger.

FIGURE 5 shows the variation in ν_2 as a function of ω using Eq. (62). Negative values of ω produce almost symmetrical variations in the opposite direction when ω is large. For small ω there is a pole at $\omega = -12/7$ where ν_2 diverges. But note that there is no growing linear mode in this case, which means that fluctuations are rapidly suppressed. So when fluctuations are growing BD gives a very small (negative) contribution to the skewness, no matter what the value of ω is.

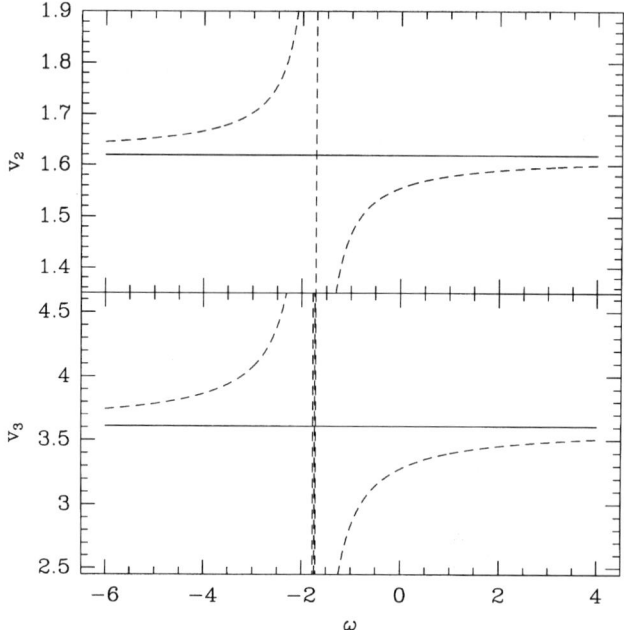

FIGURE 5. The bottom panel shows (*dashed lines*) ν_3 for a time varying gravitational constant $G = G_0 a^{-1/(1+\omega)}$, as a function of ω. The top panel shows the corresponding variation for ν_4. The continuous horizontal lines correspond to GR results $\nu_2 = 34/21$ and $\nu_3 = 682/189$.

3.3. Strongly Nonlinear Regime

FIGURE 1 shows the fully nonlinear solution for the overdensity δ as a function of the linear one δ_1. The continuous line shows the exact solution to Eq. (16). As can be seen in the FIGURE, there is a critical value of $\delta_1 = 3/2(3\pi/2)^{3/2} \simeq 1.6865$ where the nonlinear fluctuations become infinite. This corresponds to the point where the collapse occurs (see [2]).

The short-dashed lines correspond to the same exact solution in the BD model with $\omega = 10$ and $\omega = 1$.

FIGURE 6 shows δ_c, defined as the value of the linear overdensity where the corresponding nonlinear value becomes infinity, in the BD model as a function of ω.

4. DISCUSSION AND CONCLUSIONS

We have reconsidered the problem of linear and nonlinear structure formation in two different contexts that relate to observations: 1-point cumulants of large-scale density fluctuations and the epoch of formation and abundance of structures using the Press and Schechter [1] formalism. In these two contexts one can use the shear-

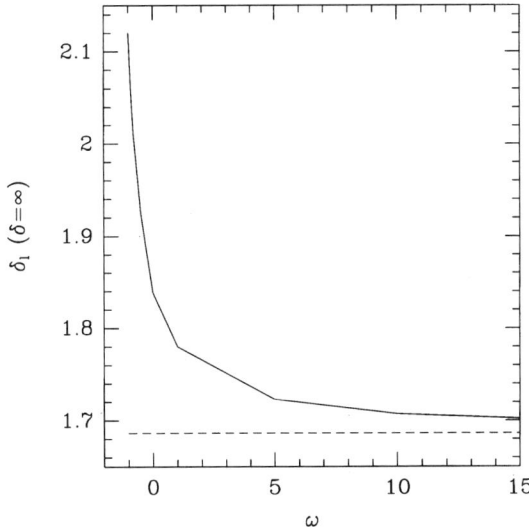

FIGURE 6. The critical value of the linear density contrast δ_c where $\delta = \infty$ as a function of ω in BD models.

free or spherical collapse (SC) model, which corresponds to the spherically symmetric (or local) dynamics. Within this approximation, we have addressed the question of how different are the predictions when using a nonstandard theory of gravity, such as BD model, or nonstandard cosmological model (e.g., a different equation of state or Hubble law).

In a separate paper [12], we present some preliminary bounds on γ and ω from observations of the skewness and kurtosis in the APM. These bounds are optimistic given the current data, but the situation is going to change rapidly, and one can hope to find much better bounds form upcoming data (such as 2DF or SDSS projects). In terms of the equation of state the bounds indicate that our Universe is neither radiation ($\gamma = 1/3$) nor vacuum dominated ($\gamma = -1$), but somewhere in between (e.g., matter dominated). In terms of the gravitational constant, the bounds on ω say that G has not changed by more than $\simeq 5\%$ from $z \simeq 1.15$, or by distances of $\simeq 400$ Mpc. Clustering at higher redshift would probe much larger scales and times. We have also shown how halo and cluster abundances and formation times could change in these nonstandard cases.

It is straightforward to combine several of the changes proposed here to explore more general situations. One could for example parametrize theories in the (γ, ω) plane, for example, different equations of state with different BD parameters, or consider the whole $(\gamma, \omega, \Omega_M, \Omega_\Lambda)$ space. This would obviously allow for a wider set of possible solutions and degeneracies. One should also consider other observational consequences of these variations, such as the age of the Universe, the growth in the radiation epoch, or the constraints from nucleosynthesis.

Rather than proposing an alternative theory of gravity or cosmological model, the aim of this paper was to show that some small deviations from the current paradigm have significant and measurable consequences for structure formation. This could eventually help explaining some of the current puzzles confronting the theory, such as the need of nonbaryonic dark matter. We have shown how current and upcoming observations of clustering and mass functions can be used to explore our assumptions and place limits on the theory of gravity at large, $\gtrsim 1h^{-1}$ Mpc, scales. This provides an interesting test for gravity as the driving force for structure formation and for our knowledge of the cosmological equation of state. A more comprehensive comparison with particular scenarios is left for future work.

ACKNOWLEDGMENTS

One of us (JAL) gratefully acknowledges financial support from the Spanish Ministry of Education, contract PB96-0384, and also Instutut d'Estudis Catalans. EG acknowledges support from CSIC, DGICYT (Spain), project PB96-0925. We would like to thank IEEC, where most of this work was carried out.

REFERENCES

1. PRESS, W.H. & P. SCHECHTER. 1974. Formation of galaxies and clusters of galaxies by self-similar gravitational condensation. Astrophys. J. **187**: 425–438.
2. PEEBLES, P.J.E. 1993. Principles of Physical Cosmology. Princeton University Press. Princenton, New Jersey.
3. WALD, R.M. 1984. General Relativity. University of Chicago Press. Chicago, IL.
4. BERNARDEAU, F. 1992. The gravity-induced quasi-Gaussian correlation hierarchy. Astrophys. J. **392**: 1–14.
5. FOSALBA, P. & E. GAZTAÑAGA. 1998. Cosmological perturbation theory and the spherical collapse model — III. The velocity divergence field and the Omega dependence. Mon. Not. R. Astr. Soc. **301**: 535–546
6. FOSALBA, P. & E. GAZTAÑAGA. 1998. Cosmological perturbation theory and the spherical collapse model — I. Gaussian initial conditions. Mon. Not. R. Astr. Soc. **301**: 503–523.
7. GAZTAÑAGA, E. & P. FOSALBA. 1998. Cosmological perturbation theory and the spherical collapse model — II. Non-Gaussian initial conditions. Mon. Not. R. Astr. Soc. **301**: 524–534.
8. BARROW, J.D. & P. PARSONS. 1997. Behavior of cosmological models with varying G. Phys. Rev. D **55**: 1906–1936.
9. WEINBERG, S. 1972. Gravitation and Cosmology. J. Wiley & Sons. New York.
10. BARROW, J.D. 1996. Time-varying G. Mon. Not. R. Astr. Soc. **282**: 1397–1406.
11. NARIAI, H. 1969. Gravitational instability in the Brans-Dicke cosmology. Prog. Theor. Phys. **42**: 544–554.
12. GAZTAÑAGA, E. & A. LOBO. 2001. Non-linear gravitational growth of large scale structures inside and outside standard cosmology. Astrophys. J. **548**: 47–59.
13. WILL, C.M. 1993. Theory and Experiment in Gravitational Physics, 2nd ed. Cambridge University Press. Cambridge.

An Attempt to Do Without Dark Matter

WILLIAM H. KINNEY AND MARTINA BRISUDOVA

Department of Astronomy, University of Florida, Gainesville, Florida 32611 USA

ABSTRACT: The discrepancy between dynamical mass measures of objects such as galaxies and the observed distribution of luminous matter in the universe is typically explained by invoking an unseen "dark matter" component. Dark matter must necessarily be nonbaryonic. We introduce a simple hypothesis to do away with the necessity for dark matter by introducing an additional non-gravitational force coupled to baryon number as a charge. We compare this hypothesis to Milgrom's modified Newtonian dynamics. The model ultimately fails when confronted with observation, but it fails in an interesting way.

KEYWORDS: Dark matter; galactic dynamics; Lensing; MOND

1. INTRODUCTION

In a dynamical sense, the universe does not behave the way we would expect based on the observed distribution of matter. Virtually every object at galactic scales and larger is being pulled on harder than can be explained simply by gravity and the distribution of matter we can see directly. The conventional (and very successful) explanation for this inconsistency is that there is matter we can't see, *dark matter*, that is responsible for the additional gravitational force. There is, however, an obvious alternative to this explanation. It could be that something about gravity changes on very large scales and is responsible for the observed dynamics of galaxies and clusters of galaxies. In fact, it would be quite surprising if general relativity, which is fairly well tested on solar system scales, were to survive unmodified when one considers scales relevant to cosmology, an extrapolation of some 14 orders of magnitude! While the standard cosmology is elegant and powerful, it is instructive (and entertaining) to consider alternatives. In this paper we describe one such alternative model. This model ultimately fails when faced with observational data, but it fails in a sufficiently interesting way that it is worth the time spent considering it.

It is not clear how far one has to go in order to construct such an alternative. Is it possible to save a reasonably "standard" cosmology in the absence of dark matter, that is, an expanding spacetime arising out of a hot big bang with a horizon and a uniform Hubble expansion? Is it possible to leave the framework of general relativity intact? Is it possible even to retain Newton's laws? One famous attempt to construct a cosmology with no dark matter is Milgrom's *Modified Newtonian Dynamics* (MOND) [1], which postulates that Newton's law of gravity fails at large distances.

Address for correspondence: William H. Kinney, Columbia Astrophysics Laboratory, 550 W. 120th Street, New York, NY 10027 USA. Voice: 212/854-0713; fax: 212/854-8121.
kinney@astro.columbia.edu

MOND introduces a fundamental acceleration a_0, such that the "true" acceleration of a body is related to the Newtonian acceleration $\vec{a}_N$ by an interpolating function μ:

$$\vec{a}_N = \mu(a/a_0)\vec{a}, \tag{1}$$

where $\mu(x)$ is taken to have the behavior

$$\mu(x) \longrightarrow \begin{cases} x & (x \ll 1) \\ 1 & (x \gg 1) \end{cases}. \tag{2}$$

This hypothesis easily explains the most compelling observational evidence for the existence of dark matter, which is the flatness of galactic rotation curves. It is a simple matter to see that the MOND hypothesis gives flat rotation curves as a natural consequence. Taking the acceleration of a body in a circular orbit of radius r,

$$a = \left(\frac{v^2}{r}\right) \tag{3}$$

we write the relation to the Newtonian gravitational force in the MOND limit $\mu(x) \to x$ as

$$a_N \to \frac{a^2}{a_0} = \frac{v^4}{a_0 r^2} = \frac{GM}{r^2}. \tag{4}$$

The orbital velocity v is then just a combination of constants:

$$v^4 = GMa_0 = \text{const}, \tag{5}$$

so that the observed flat rotation curves of galaxies at large radii are a natural consequence of MOND. The constant a_0 can be determined from rotation curve data, with the "standard" value being

$$a_0 = 1.1 \times 10^{-10} \, h_{70} \, \text{m/s}^2, \tag{6}$$

where h_{70} is the value of the Hubble constant H_0 in units of 70 km/sec/Mpc. MOND has in fact been shown to work very well as an explanation for the observed rotation curves of galaxies [2], [3]. It is, however, a radical hypothesis. Consider the force between two objects of mass m and mass M, respectively, in the MOND limit:

$$F_{\text{MOND}} = m\sqrt{a_0 a_N}$$
$$\propto m\sqrt{M}. \tag{7}$$

We see that one must abandon even Newton's rule that every action has an equal and opposite reaction. Surely this is too high a price to pay for explaining even a large body of astrophysical data! We wish to construct a less sweeping alternative, one which leaves as much standard physics intact as possible: we wish to get rid of the bath water, but still keep the baby.

2. A MORE PALATABLE ALTERNATIVE TO MOND?

In this section we construct an alternative to MOND which provides a good fit to cosmological observations without the requirement either for dark matter or for radical revisions to basic physics. As discussed in the previous section, this model is motivated by the observation that the standard cosmology requires gravity to be the only relevant force operating over a span of 14 or more orders of magnitude in scale, which would make it unique among fundamental forces. It is certainly a possibility that a new, nongravitational force becomes dominant on very large scales,

$$V(r) = -\frac{GMm}{r} + V_1(r), \quad (8)$$

where the potential due to the nongravitational component is labeled V_1. Such a new interaction must couple to a charge of some kind, and we note that standard particle physics provides an excellent candidate: baryon number. An interaction mediated, for example, by a scalar particle coupled to baryon number would act as a universal attractive force similar to gravity, but with distinct scales and couplings. A naive model of a scalar-mediated force, however, will not work for our purpose since by gauss' law, the force must fall of as $1/r^2$ and if the new force is to dominate over gravity at large scales it must do so at *all* scales. We postulate instead a force which falls off as $1/r$, with potential

$$V_1(r) = \alpha \ln(r), \quad (9)$$

where the constant α is given in terms of the baryon numbers b of the interacting bodies and a fundamental mass scale Λ,

$$\alpha = \Lambda b_1 b_2. \quad (10)$$

We have no fundamental model which creates this behavior, but simply take it as a phenomenological guess. Such a logarithmic potential leads naturally to flat rotation curves. The acceleration of a body in orbit around a central mass M can be written

$$a(r) = \frac{GM}{r^2} + \frac{\alpha}{mr} = \frac{\alpha}{mr}\left[1 + \left(\frac{r_0}{r}\right)\right], \quad (11)$$

where the fundamental length scale r_0 is defined as

$$r_0 \equiv \frac{GMm}{\alpha}. \quad (12)$$

Then, for $r \gg r_0$,

$$a \to \frac{\alpha}{mr} = \frac{v^2}{r}. \tag{13}$$

and we have flat rotation curves

$$v^2 = \frac{\alpha}{m} = \text{const.} \tag{14}$$

To be precise, the rotation velocity is constant as long as the baryon number-to-mass ratio of the galactic matter is constant, since

$$\alpha = \Lambda b_1 b_2 = \Lambda \left(\frac{b_1}{M}\right)\left(\frac{b_2}{m}\right) Mm \simeq \frac{\Lambda M m}{m_p^2} \tag{15}$$

where m_p is the proton mass. This expression is valid up to the proton/neutron mass ratio. The rotation velocity of a body about a galaxy can then be expressed as a ratio of mass scales,

$$v^2 \simeq \frac{\Lambda M_{\text{gal}}}{m_p^2}. \tag{16}$$

This model can be cast in a form very similar to MOND, in which the true acceleration and the Newtonian acceleration are related through an interpolating function,

$$a_N = \frac{1}{1 + (r/r_0)} a = \mu(a/a_0) a. \tag{17}$$

where the asymptotic acceleration a_0 is given by

$$a_0 = \frac{\alpha}{mr_0}, \tag{18}$$

and the interpolating function $\mu(x)$ is given by

$$\mu(x) = \frac{\sqrt{1+4x}-1}{\sqrt{1+4x}+1} \tag{19}$$

It is important to note, however, that this model is *not* equivalent to MOND, since the asymptotic acceleration a_0 is not a fundamental constant, but varies from galaxy to galaxy:

$$a_0 = \frac{\alpha}{mr_0} \simeq \left(\frac{\Lambda^2}{Gm_p^2}\right) M_{\text{gal}}, \tag{20}$$

Instead, the radius r_0 is a universal constant,

$$r_0 = \frac{GMm}{\alpha} \simeq \frac{Gm_p^2}{\Lambda}, \qquad (21)$$

To explain galactic rotation curves, the fundamental radius r_0 must be of order 10 kpc or so. (In the next section we derive an independent estimate of r_0 based on X-ray observations of galaxy clusters.) Unlike the breakdown of Newtonian physics which occurs in MOND, $F_{\text{MOND}} \propto m\sqrt{M}$, Newton's third law is preserved,

$$F \propto \frac{b_1 b_2}{r} \sim \frac{Mm}{r}, \qquad (22)$$

Our hypothesis leaves virtually all of standard physics intact.

Flat rotation curves, however, are but one class of a host of astrophysical observations which must be fit with such a model. In the next section, we describe the confrontation of our hypothesis with observation. Remarkably, the model fits a variety of independent constraints, although we ultimately find the model lacking and reject it as a plausible scenario.

3. CONFRONTATION WITH OBSERVATION

Galactic rotation curves are just the beginning. Our model provides a tidy, minimalist explanation for flat rotation curves without the requirement for nonbaryonic dark matter. Such an additional force, however, will dominate over gravity on all scales $r > r_0$ and will therefore be subject to a large number of observational constraints over a range of scales, including galactic scales, cluster scales, large-scale structure, and the universe as a whole:

- Galaxies
    Galactic masses
    Tully-Fisher relation
- Clusters
    Dynamical mass measures
    X-ray mass measures
    Lensing
- Large-scale structure
    Silk damping and primordial perturbations
- Cosmology

We include cosmology in the list because general relativity is left intact in our model, and the Friedmann–Robertson–Walker universe is still a viable cosmological model. The extra force simply adds a term to the stress-energy in the Einstein field equations:

$$G_{\mu\nu} = 8\pi G [T_{\mu\nu}^0 + \delta T_{\mu\nu}], \qquad (23)$$

where $T^0_{\mu\nu}$ is the usual stress-energy of the baryons, and $\delta T_{\mu\nu}$ is the contribution from the interaction. A fully relativistic model is required to calculate $\delta T_{\mu\nu}$, and we do not consider this question further here. Large-scale structure provides a strong constraint on a baryon-only cosmology as well, since photon diffusion (Silk damping) erases primordial fluctuations in the baryon fluid on scales smaller than the horizon at early times. We also leave this important question hanging, and instead concentrate on galactic and cluster scales. We find that these smaller scales alone are sufficient to rule out the model.

We first consider galactic scales. The first and simplest question we can ask is what is the typical mass of a galaxy? Note that we are assuming that galaxies are made up entirely of baryons, so galactic masses should be neither too large nor too small relative to their luminosities. We can write the mass of the galaxy in terms of the rotation velocity v_c as

$$M_{gal} = \frac{r_0 v_c^2}{G} = 1.1 \times 10^{10} M_\odot \left(\frac{r_0}{1 \text{ kpc}}\right)\left(\frac{v_c}{220 \text{ km/s}}\right)^2 \qquad (24)$$

This figure is nicely consistent with typical galaxy luminosities of $\mathcal{L} \sim$ few $\times$ 10^{10} $\mathcal{L}_\odot$, so the assumption of a baryon-only halo works very well in this model. We can gain further constraints on galactic mass-to-light ratios by use of the Tully–Fisher relation, an empirical relation between a galaxy's luminosity and rotation velocity,

$$\mathcal{L} \propto v_c^4. \qquad (25)$$

The simplest and most physical assumption for a baryon-only halo is that a galaxy's luminosity is somewhere close to proportional to its mass, $\mathcal{L} \propto M_{gal}$. That is, all baryons are roughly equally luminous. This is in fact a natural consequence of MOND, since MOND predicts $M_{gal} \propto v_c^4$ and therefore $M_{gal} \propto \mathcal{L}_{gal}$ from the Tully–Fisher relation. Explaining flat rotation curves with an extra force, however, requires a strong variation of luminosity with mass, since $M_{gal} \propto v_c^2$, and, from Tully–Fisher,

$$\mathcal{L}_{gal} \propto v_c^4 \propto M_{gal}^2. \qquad (26)$$

While it is difficult to understand why a galaxy might have such a relationship between mass and luminosity, neither is it ruled out by observation. We will simply take it as a prediction of the model.

What about larger scales? The rich data available on galaxy clusters provide several useful constraints on the model. We can divide observations of galaxy clusters into three general classes. First, observations at X-ray wavelengths and measurements of the Sunyaev–Zel'dovich (SZ) effect in clusters provide a direct view of the baryons in the cluster. They allow us to "count" the baryons directly. Dynamical measures such as velocity dispersion or mass measurements based on assumptions of hydrostatic equilibrium probe the binding energy of the cluster. With the assumption of Newtonian gravity, these measures actually "weigh" the cluster, but with the assumption that a nongravitational force dominates on cluster scales, dynamical measurements in fact probe the form of the force. We show below that combining "baryon counting" measurements and dynamical measurements allows us to fix r_0. The third class of measurements is gravitational lensing, which we assume (unlike the dynamical measurements) to be governed by general relativity and produce a direct measure of the mass of the cluster. We save a discussion of lensing for last, because it rules out the model.

With the assumption that Newtonian gravitation is the dominant force binding a galaxy cluster together, a combination of X-ray measurements and dynamical measurements is generally used to fix the "baryon fraction" f_B of the cluster. We will assume a typical value for f_B of [4]

$$f_B = 0.06 \, h^{-3/2} \tag{27}$$

In the standard picture, such a determination of the cluster baryon fraction is powerful evidence for the existence of dark matter. Since we wish to do without dark matter, we shall assume that the *true* baryon fraction is unity $f_B = 1$ and that the apparent baryon fraction of galaxy clusters is an artifact of the additional force which dominates over gravity at cluster scales. This is sufficient to fix the fundamental length r_0. Consider a cluster with velocity dispersion σ and core radius and density r_c and ρ_c. In a simple model of such a cluster [5], the density as a function of radius can be written

$$a\rho(r) = \frac{3\rho_c \sigma^2 r}{r_c^2}, \tag{28}$$

where a is the acceleration. In our model of an extra force, the true acceleration is related to the Newtonian acceleration a_N as

$$a = \left(\frac{r}{r_0}\right) a_N, \tag{29}$$

so the true density of the cluster is related to the apparent Newtonian density ρ_N as

$$\rho(r) = \left(\frac{r_0}{r}\right) \rho_N(r). \tag{30}$$

In other words, assuming Newtonian gravity causes us to *underestimate* the baryon fraction in a cluster at a radius r by a factor of (r/r_0). Taking the characteristic radius to be a typical cluster core radius $r_c \sim 0.2 \, h^{-1}$ Mpc [5], we can relate the apparent baryon fraction f_B^N to the true baryon fraction as

$$f_B \equiv 1 = \left(\frac{r_c}{r_0}\right) f_N^B. \tag{31}$$

We then have an estimate for the scale r_0 that agrees amazingly well with what we would expect from galactic dynamics,

$$r_0 = 1.2 \, h^{-5/2} \text{ kpc} \simeq 4.3 \text{ kpc} \ (h = 0.6). \tag{32}$$

Our assumption of a baryon-only universe and an extra force works better than we perhaps have any right to expect, explaining (at least roughly) the independent phenomena of galactic disk dynamics and galaxy cluster dynamics within a single simple framework.

Unfortunately, our run of luck ends when we consider gravitational lensing by clusters. Since our hypothetical force couples to baryon number as a charge, we expect it to interact with photons only via loop effects. Therefore gravitational lensing by the cluster will measure only the gravitational potential, or, equivalently, the actual mass of the cluster. We therefore expect lensing mass estimates to be significantly lower than dynamical mass estimates, "underestimating" the cluster mass by a factor of $1/f_B^N \simeq 10$. In fact, this is not so. In a survey of lensing mass estimates of clusters, Wu *et al.* find that weak lensing mass estimates agree well with both mass estimates determined from velocity dispersion and from hydrostatic equilibrium [6]. Wu *et al.* in fact find that mass estimates from strong lensing tend to *overestimate* the the cluster masses by a significant amount. There is no evidence for systematically low mass estimates that would be required if our extra force model were correct. Lensing kills the model.

CONCLUSIONS

We have presented a hypothesis for a way to construct a universe with no dark matter, in which an extra force with coupling to baryon number dominates over gravitation on scales larger than a few kpc. A logarithmic potential results in naturally flat galactic rotation curves and very consistently explains the apparent baryon fraction of galaxy clusters $f_B < 1$ as an artifact of the new force. Gravitational lensing, however, can't be fooled, since it measures the actual mass of galaxy clusters, regardless of the force law responsible for their dynamics. The model fails.

So why bother even talking about such an alternative model? First, it brings to light a general fact about alternatives to dark matter. Whatever our "extra" force might be, it must couple to photons (lensing) in exactly the same way it couples to ordinary matter (dynamics). In other words, it must act like gravity. And anything that is coupled to matter and radiation exactly like gravity must, in fact, *be* gravity. This is suggestive of the conclusion that there is simply no way to do away with dark matter without significant modifications to general relativity itself.

Finally, although our attempt to invoke a new force as a way to eliminate dark matter ultimately fails, such additional forces could still be of cosmological interest. A particularly interesting question is what effect a weak long-range force would have on the evolution of the universe as a whole, independent of any assumptions of the composition of the matter in the universe.

REFERENCES

1. MILGROM, M. 1983. A modification of the Newtonian dynamics as a possible alternative to the hidden mass hypothesis. Astrophys. J. **270:** 365–370.

2. BEGEMAN, K.G., A.H. BROEILS & R.H. SANDERS. 1991. Extended rotation curves of spiral galaxies: dark haloes and modified dynamics. Mon. Not. Roy. Astr. Soc. **249:** 523–537.
3. SANDERS, R.H. 1996. The published extended rotation curves of spiral galaxies: confrontation with modified dynamics. Astrophys. J. **473:** 117–129.
4. EVRARD, A.E. 1997. The intracluster gas fraction in X-ray clusters — Constraints on the clustered mass density. Mon. Not. Roy. Astr. Soc. **292:** 289–297.
5. PEEBLES, P.J.E. 1993. Principles of Physical Cosmology: 434-436. Princeton University Press, Princeton, NJ.
6. WU, X-P., T. CHIUEH, L. FANG & Y-J. XUE. 1998. A comparison of different cluster mass estimates: consistency or discrepancy? Mon. Not. Roy Astr. Soc. **301:** 861–871.

Emergence of Anomalous Distributions in Disordered Systems

K. A. MUTTALIB[a] AND P. WÖLFLE[b]

[a]Department of Physics, University of Florida, P.O. Box 118440, Gainesville, Florida 32611-8440, USA

[b]Institut für Theorie der Kondensierten Materie, Universität Karlsruhe, D-76128 Karlsruhe, Germany

> ABSTRACT: The present day non-gaussian distribution of mass density of the universe evolved from an initial gaussian distribution in the presence of nonlinear interactions. We discuss an analog in disordered condensed matter system where increasing the disorder changes the distribution of conductances from a gaussian at weak disorder to a log-normal at strong disorder. The highly asymmetric "one-sided" log-normal distribution in the intermediate crossover regime can be understood as a simple hybrid of these two limiting distributions.
>
> KEYWORDS: Conductance distributions; Disordered systems; Metal-insulator transitions

1. INTRODUCTION

It is believed that the distribution of mass density in the early universe started as gaussian, but it evolved towards the present day non-gaussian distribution due to the nonlinearities associated with cluster formation in the presence of gravity. One can equivalently study the present day distribution as a function of scale or size, which is gaussian at large scales but has long tails at small scales. In disordered condensed matter systems, a similar situation occurs when one starts with a gaussian distribution for an observable and there exists an *experimental knob* that can be turned (for example, changing the number of scattering centers in a given sample) to change the distribution. In such a case one can in principle study in a systematic and controlled way how a gaussian distribution suddenly starts "growing a tail" as one keeps on turning the experimental knob. An even more interesting situation occurs if the system undergoes a phase transition while the knob is being turned, and leads to a nontrivial limiting distribution on the other side of the transition. Then the gaussian distribution not only develops a tail, but has to change dramatically through the phase transition. In this review we will consider such an example.

In particular, the distribution on the other side of the transition we consider is lognormal, which is a gaussian distribution for the logarithm of the same variable. In

---

Address for correspondence: Khandker Muttalib, Department of Physics, University of Florida, P.O. Box 118440, Gainesville, FL 32611-8440 USA. Voice: 352/392-6699; fax: 352/846-0295.

muttalib@phys.ufl.edu

other words, the distribution changes from gaussian to log-normal as a an experimental parameter is changed. At the intermediate crossover or transition regime, the distribution is of a novel type which is highly asymmetric with diverging moments and can not be studied within a perturbative scheme. Nevertheless, it turns out that the novel distribution can be understood as a simple hybrid of the two limiting distributions.

2. THE CONDUCTANCE

We will discuss a system where noninteracting electrons move in a periodic lattice with random scattering centers and experiments measure electrical transport properties of the system. The density of scattering centers can be increased systematically by, for example, bombarding the pure sample with a specific dose of neutrons or alpha particles to create defects. Thus there exists an experimental knob to change the amount of disorder in the system. This apparently simple system shows a surprisingly rich behavior. Without the randomness, the system behaves as a perfect conductor with infinite conductivity. This is because the solution of the Schrödinger equation for a perfect periodic lattice is a Bloch wave, which does not scatter from the lattice points. In the presence of "weak" disorder, the system is a metal; because of the repeated scattering from the random scattering centers, the electrons "diffuse" through the system as in a random walk problem, giving rise to a finite conductivity. For sufficiently large disorder, the electrons get trapped in one of the deep minima of the random potential, and coherent backscattering from nearby scattering centers lead to an exponentially localized wavefunction that does not carry any current; the system becomes an insulator [1]. In three dimensions, there exists a critical disorder such that the system undergoes a second-order phase transition from the metallic to the insulating behavior as the disorder is increased through the critical disorder [2].

We can imagine a disordered conductor to be described by a scattering matrix S, which relates the incoming flux of electrons to the outgoing flux, $S\binom{I}{I'} = \binom{O}{O'}$, where I, O, I', O' describe the wave amplitudes on the left and right, respectively. In terms of the transmission and reflection matrices t, t' and r, r' the S matrix can be written in the form

$$S = \begin{pmatrix} t & r \\ r' & t' \end{pmatrix}.$$

Then the dimensionless conductance $g = Tr(tt^\dagger)$ is a measure of the capacity of the system to carry current from one side of the system to the other when a voltage is applied [3]. In this dimensionless form, a metallic sample of finite size will correspond to having $g \gg 1$ (weak disorder), and an insulator will correspond to $g \ll 1$ (strong disorder). The metal-insulator transition occurs at the critical conductance $g \sim 1$ (critical disorder).

3. DISTRIBUTION OF CONDUCTANCES: GAUSSIAN VERSUS LOG-NORMAL

If we take several different samples with the *same disorder*, which means the same density of scattering centers but *different realizations* of the randomness, then the measured values of the conductances will not be identical, giving rise to a distribution of conductances $P(g)$. It is known from perturbation theory and experiments that in the metallic regime, the distribution $P(g)$ is gaussian. Typically a macroscopic system is *self-averaging*; a single measurement corresponds to an already ensemble averaged value. This is because one can think of the current as flowing along many parallel paths, each characterized by a partial conductance. To the extent that these conductances are statistically independent, the total conductance of the sample, which is the sum of the partial conductances, has a gaussian distribution. For such systems, it turns out, somewhat surprisingly, that the variance of the distribution is a *universal* quantity of order unity (2/15 for a quasi one-dimensional system considered here), independent of the size of the system or the mean value of the conductance, as long as the system remains metallic. This is known as the universal conductance fluctuation [4]. This means that as one turns the disorder knob, the mean conductance decreases with increasing disorder, but the distribution $P(g)$ around that mean remains gaussian and the width of the distribution does not get broader as one would have naively expected. Clearly, changing disorder in this regime does not lead to growing tails to the gaussian distribution.

On the other hand, the distribution of conductances is known to be log-normal in the insulating regime [5]. This is because the electron wavefunctions in this regime are exponentially localized within a region of the size of the so-called localization length. In this case, electron transport is taking place predominantly via a single wavefunction spanning the system, all the other states giving an exponentially smaller contribution. The conductance is thus determined by the statistically fluctuating (from sample to sample) localization length of the dominating wavefunction. Assuming a Gaussian distribution of the inverse localization length, it follows that the logarithm of the conductance will be gaussian distributed, which is a log-normal distribution for $P(g)$. This is a highly asymmetric distribution with a very long tail. The implication is that although $P(g)$ remained gaussian and even the width did not change with increasing disorder for small disorder, once the disorder becomes sufficiently strong, $P(g)$ changes from a gaussian on the metallic side to a log-normal on the insulating side. It must be true then that somewhere near the critical disorder, the gaussian suddenly starts to grow tails.

4. TRANSITION FROM GAUSSIAN TO LOG-NORMAL DISTRIBUTION

If the critical distribution is not symmetric and has long tails, then the usual description of the phase transition in terms of the mean (or the most probable) conductance is no longer adequate because the tails may dominate the physical properties. In this case the phase transition should be described in terms of a transition between two distributions. Then we need to ask: what is the nature of the distribution near the critical regime? How does it emerge from either the gaussian or the log-normal distributions as one turns the disorder knob one way or the other?

Clearly, perturbation theory in disorder (or $1/g$) is inadequate to understand the anomalous distribution near the transition, because significant changes in the gaussian occur only at strong disorder near the critical value of $g \sim 1$. Moreover, the nth moment of the log-normal distribution diverge faster than e^{n^2}, which means that the distribution cannot be uniquely reconstructed from the moments. If the moments of the critical distribution also diverge similarly, then one needs to consider the full distribution $P(g)$ directly rather than trying to extract it from the moments. Indeed, a combination of perturbation theory in $d = 2 + \epsilon$ dimensions [6] and an attempt to reconstruct $P(g)$ from the resulting diverging moments lead to a prediction for the critical distribution having a gaussian head (centered at $1/\epsilon$) and symmetric power law tails [7]. This picture is not even remotely close to the experimental [8] or numerical evidences [9], which show a highly asymmetric critical distribution with log-normal tail on the insulating side and a very sharp cutoff on the metallic side at $g \sim 1$. A possible reason for this failure is that the critical conductance in the $2 + \epsilon$ dimension happens to be at $g_c \sim 1/\epsilon$ which, for the perturbation theory to be valid, must be at $\epsilon \ll 1$, that is, $g_c \gg 1$. This is deep in the metallic regime, and fails to capture the actual critical regime at $g_c \sim 1$ which is beyond the perturbative limit.

We have developed a novel nonperturbative method to study the distribution $P(g)$ directly without having to compute the moments. The method is valid only in quasi-one dimensions ("thick quantum wire"), where localization of the wavefunction along the direction transverse to the current is not fully taken into account. This limits us to the situation where there is only a crossover between the metallic and insulating regimes and no true phase transition as in three dimensions. However, since the localization length diverges near the critical point, even in quasi–one-dimension the crossover regime gives a qualitatively good description of the critical distribution. In fact, our predictions at the middle of the crossover agree quite well with those obtained numerically for the true phase transition in three dimensions.

The method is based on a Fokker–Planck equation for the joint probability distribution (jpd) of the transmission eigenvalues for an N channel wire of length L, which describes the "evolution" of the distribution as the length is increased [10]. The parameter $\Gamma = Nl/L$ where $l \ll L$ is the mean free path plays the role of the strength of disorder. (The thermodynamic limit is taken as $N \to \infty$, $L \to \infty$, keeping the ratio Γ to be finite.) Conductance $g = Tr(tt^{\dagger})$ is directly related to the real, nonnegative transmission eigenvalues λ_i by the relation

$$g = \sum_{i=1}^{N} \frac{1}{1 + \lambda_i},$$

where a metal consists of a fraction l/L of λ_i, each much smaller than unity (these are the "open" channels), so that the conductance is $g = Nl/L = \Gamma$; an insulator will consist of all $\lambda_i \gg 1$, giving rise to a conductance $g \ll 1$. $P(g)$ can then be written down as an N-dimensional integral of the jpd of the transmission eigenvalues λ_i which are real and nonnegative [11]. By writing the jpd as the exponential of an effective Hamiltonian, one can then map the problem on to an electrostatic problem of N charges on a line (from 0 to ∞) with two-particle repulsive interactions and a single particle confinement potential. The electrostatic problem is then solved within a variational scheme [12]. Note that the distribution for all disorder in this framework turns out to be a function of only one parameter, Γ, with $\Gamma \gg 1$ corresponding to a metal and $\Gamma \ll 1$ corresponding to an insulator. The crossover occurs at $\Gamma \sim 1$.

It turns out that understanding how tails *grow* in the gaussian distribution is far more complicated than understanding how the tails in the log-normal distribution get *cut off* as one approaches the critical regime from the insulating side. In our framework this is because the metallic distribution ($g \gg 1$) involves the jpd of small $\lambda_i \ll 1$ which are strongly correlated due to their mutual repulsive interaction, while the insulating distribution ($g \ll 1$) involves only the smallest one or two eigenvalues which are exponentially separated from the rest. In terms of the physical picture presented above, conduction in the metallic regime ($g > 1$) involves different current paths coupled in parallel. For samples of this type, the distribution will be gaussian, even though only the tail of the distribution may fall into the regime $g > 1$. As soon as one current path (transmission channel) is dominant ($g < 1$), the distribution changes to log-normal. As a consequence, the transition between the two regimes $g < 1$ and $g > 1$ in $P(g)$ turns out to be remarkably sharp. Our results predict [12] that in the insulating regime (in the absence of time reversal symmetry) the distribution has the form

$$P[\ln(g)] \approx \begin{cases} \sqrt{\dfrac{x_1 \sinh 2x_1}{1-g}} e^{-\Gamma x_1^2}, & g < 1, \\ \sqrt{2}g e^{-a(g-1)^2}, & g < 1, \end{cases}$$

where $x_1 = \cosh^{-1}(1/\sqrt{g})$ and a depends on the disorder parameter Γ.

FIGURE 1 shows plots of $P[\ln(g)]$ as a function of $\ln(g)$ given by the above equation for four values of Γ. In the deeply insulating regime $\Gamma \ll 1$, the distribution is log-normal with peak at $\langle \ln(g) \rangle = 1/\Gamma$. As the disorder is decreased, Γ increases and the peak of the log-normal distribution shifts towards larger g with increasing g. For these cases, the distribution gets the $g > 1$ part of its tail cut off by a gaussian. This gives rise to a highly asymmetric distribution which eventually leads to a novel "one-sided log-normal" distribution when the peak of the log-normal distribution shifts to $g = 1$. The Gaussian cut off at $g = 1$ is indicative of the fact that $g = 1$ requires one "open" channel for the current, corresponding to the first transmission eigenvalue $\lambda_1 \ll 1$. However, the repulsive interaction between the eigenvalues forces the next one exponentially far for $\Gamma \ll 1$, leading to a sharp cutoff for the distribution beyond $g > 1$.

"One-sided log-normal" distributions predicted above have already been observed (at least with the same qualitative features) in numerical investigations of the crossover/transition regimes of integer Quantum Hall systems in quasi–one, as well as, –two dimensions and Anderson metal-insulator transitions in three dimensions

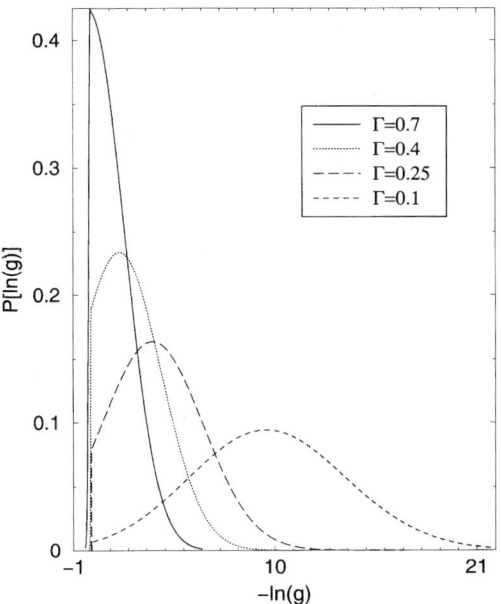

FIGURE 1. $P[\ln(g)]$ as a function of $\ln(g)$ for four different values of Γ in the insulating regime $\Gamma < 1$. *Dashed line* corresponds to $\Gamma = 0.1$; *long dashed line,* $\Gamma = 0.25$; *dotted line,* $\Gamma = 0.4$; and *solid line,* $\Gamma = 0.7$ — which is close to the crossover regime.

[9]. The details of how the log-normal tail gets cut off at $g \sim 1$ have, however, not yet been studied numerically. It would be interesting to see if such highly asymmetric hybrid distributions appear generically in crossover situations between two different limiting distributions.

The present day non-gaussian distribution of mass density in the universe look remarkably close to a log-normal distribution [13]. Is it possible that the distribution is actually a hybrid of two simple limiting distributions in some intermediate regime of its time evolution?

REFERENCES

1. ANDERSON, P.W. 1958. Phys. Rev. **109:** 1492.
2. ABRAHAMS, E., P.W. ANDERSON, D.C. LICCIARDELLO & T.V. RAMAKRISHNAN. 1979. Phys. Rev. Lett. **42:** 673.
3. LANDAUER, R. 1970. Philos. Mag. **21:** 863.
4. ALTSHULER, B.L. 1985. JETP Lett. **41:** 648; P.A. LEE & A. D. STONE. 1985. Phys. Rev. Lett. **5:** 1622.
5. ANDERSON, P.W., D.J. THOULESS, E. ABRAHAMS & D.S. FISHER. 1980. Phys. Rev. B **22:** 3519.
6. ALTSHULER, B.L., V.E. KRAVTSOV & I.V. LERNER. 1986. Sov. Phys. JETP **64:** 1352; ALTSHULER, B.L., V.E. KRAVTSOV & I.V. LERNER. 1989. Phys. Lett. A **134:** 488.

7. SHAPIRO, B. 1990. Phys. Rev. Lett. **65:** 1510.
8. POIRIER, W., D. MAILLY & M. SANQUER. 1999. Phys. Rev. B **59:** 10856.
9. PLEROU, V. & Z. WANG. 1998. Phy. Rev. B **58:** 1967; P. MARKOS. 1999. Phys. Rev. Lett. **83:** 588; T. OHTSUKI, K. SLEVIN & T. KAWARABAYASHI. eprint cond-mat/9809221.
10. DOROKHOV, O.N. 1982. JETP Lett. **36:** 318; P. A. MELLO, P. PEREYRA & N. KUMAR. 1988. Ann. Phys. (NY) **181:** 290.
11. BEENAKKER, C.W.J. & B. REJAEI. 1993. Phys. Rev. Lett. **71:** 3689.
12. MUTTALIB, K.A. & P.WÖLFLE. 1999. Phys. Rev. Lett. **83:** 3013; P.WÖLFLE & K.A. MUTTALIB. Ann. Physik. 1999. Ann. Phys. (Leipzig) **8:** 753.
13. BERNARDEAU, F. & L. KOFMAN. 1995. Astrophys. J. **443:** 479.

The Onset of Nonlinearity in Cosmological Structure

J.N. FRY[a] AND CHUNG-PEI MA[b]

[a]*Department of Physics, University of Florida, Gainesville, Florida 32611-8440, USA*

[b]*Department of Physics and Astronomy, University of Pennsylvania, Philadelphia, Pennsylvania 19104, USA*

> ABSTRACT: We discuss the progression of growth of cosmological structure, from the quasilinear evolution of nearly Gaussian fluctuations on large scales into highly non-Gaussian, strongly nonlinear structure on small scales. A systematic development in perturbation theory describes the first departures from homogeneity but fails to reproduce the fully nonlinear results. Physical insight, conceptual models, and symmetries are useful in the strong clustering regime. A phenomenological model with input information from the quasilinear regime provides enticing results for the strongly nonlinear regime.
>
> KEYWORDS: Cosmology; Large-scale structure of Universe

1. INTRODUCTION

On large scales the universe is nearly homogeneous and isotropic, but on any finite scale irregularities are apparent. Quantitatively, the variance of mass density fluctuations smoothed over a scale R that contains on average mass $\overline{M}$, $\langle (\Delta M/\overline{M})^2 \rangle = \bar{\xi}(R)$, becomes small on large scales, and higher irreducible moments vanish more quickly than $\bar{\xi}^{p/2}$ in the limit of large R, so that on large scales the distribution becomes more and more precisely Gaussian. On small scales, the opposite occurs: observations of galaxy correlations find a hierarchy of higher-order irreducible correlations that grow as a power of scale, $\xi(r) \sim (r/r_0)^{-\gamma}$ with $r_0 \approx 5\,h^{-1}$ Mpc, $\gamma \approx 1.8$; $\xi_3 \sim r^{-\gamma_3}$, with $\gamma_3 = 3.64 \pm 0.03 \approx 2\gamma_2$; and $\xi_4 \sim r^{-\gamma_4}$, with $\gamma_4 = 5.40 \pm 0.20 \approx 3\gamma$ [1]–[3]. That the connected p-point function scales as $p-1$ factors of the two-point function suggests an association between the general function and so-called tree graphs (p points or nodes connected by $p-1$ links with no cycles),

$$\xi_p = Q_p \sum_{\text{trees}} \xi(x_1)...\xi(x_{p-1}), \tag{1}$$

where the arguments of the product of two-point functions correspond to links of the corresponding graph [4]. This pattern has been found to be consistent with observa-

---

Address for correspondence: J.N. Fry, Department of Physics, New Physics Bldg. University of Florida, Gainesville, Florida 32611-8440, USA. Voice: 352/392-6692; fax: 352/392-5339.

fry@phys.ufl.edu

cpma@physics.upenn.edu

tions up to $p = 10$ [5]–[7]. Volume-averaged correlations inherit this pattern, $\bar{\xi}_p = S_p \bar{\xi}^{p-1}$, where $\bar{\xi} = \bar{\xi}_2$, $\bar{\xi}_p = \int_V d^3x_1 \ldots d^3x_p \xi_p(x_1, \ldots, x_p)/V^p$, and S_p is of order $p^{p-2} Q_p$. Interestingly, this pattern arises both for the weakly nonlinear evolution of nearly Gaussian initial fluctuations in perturbation theory, and also for self-similar scale-invariant fluctuations in the strongly nonlinear regime, although for different reasons.

In this article, we discuss the nonlinear behavior of density correlations, in the quasilinear regime, where perturbation theory is useful, and also in the strongly nonlinear regime. Perturbation theory on large scales describes the leading nonvanishing contributions, first departures from a Gaussian, and higher-order corrections, but perturbation theory eventually breaks down for $\bar{\xi} \approx 1$. In the nonlinear regime, solution is difficult, but conceptual models, physical insight, and symmetry are found to be useful. A recent development, a phenomenological construction based on the profile shapes, mass distribution, and correlations of massive halos, appears to reproduce qualitatively and quantitatively both large and small scale, weak and strong clustering regimes.

2. GROWTH OF STRUCTURE IN THE QUASILINEAR REGIME

In an expanding universe with cosmological scale factor $a(t)$, with comoving position x (related to proper position r by $r = a(t)x$), the evolution of the density contrast $\delta(x)$, defined as $\rho(x) = \bar{\rho}[1 + \delta(x)]$ and the peculiar velocity $v = a\dot{x}$ are determined by the equation of continuity,

$$\frac{\partial \delta}{\partial t} + \frac{1}{a}\nabla \cdot [(1+\delta)v] = 0, \qquad (2)$$

and the Euler equation in an expanding universe,

$$\frac{1}{a}\frac{\partial}{\partial t}(av) + \frac{1}{a}(v \cdot \nabla)v = -\frac{1}{a}\nabla\phi, \qquad (3)$$

where the gravitational potential follows from the Poisson equation,

$$\frac{1}{a^2}\nabla^2 \phi = 4\pi G \bar{\rho} \delta. \qquad (4)$$

These combine to give a single second-order equation for the evolution of the density contrast[8],

$$\frac{\partial^2 \delta}{\partial t^2} + \frac{2\dot{a}}{a}\frac{\partial \delta}{\partial t} - 4\pi G\bar{\rho}\delta = 4\pi G\bar{\rho}\delta^2 + \frac{1}{a^2}\nabla\phi \cdot \nabla\delta + \frac{1}{a^2}\nabla_i\nabla_j[(1+\delta)v^iv^j], \qquad (5)$$

Equation (5) is nonlinear, but on large scales, where the amplitude of fluctuations is small, a systematic perturbative expansion is feasible. To first order we ignore the terms on the right-hand side of Eq. (5),

$$\frac{\partial^2 \delta^{(1)}}{\partial t^2} + \frac{2\dot{a}}{a}\frac{\partial \delta^{(1)}}{\partial t} - 4\pi G \bar{\rho}\delta^{(1)} = 0. \qquad (6)$$

At linear order, no spatial derivatives appear, and the solution is of the form $\delta(x, t) = D(t)\delta_0(x)$, where $\delta_0(x) = \delta(x, t_0)$ is the initial condition at some early time t_0, and $D(t)$ is the growing solution of Eq. (6). In the matter dominated, Einstein–de Sitter case, matter density $\Omega_m = 1$ and no curvature or cosmological constant, the solution is $D(t) \sim t^{2/3} \sim a(t)$. With a power law initial power spectrum, $P_0(k) \sim k^n$, the two-point function is $\xi(x, t) \sim a(t)^2/x^{(3+n)}$. This, plus the Gaussian initial conditions expected to be inherited from the early universe, is enough to determine the behavior of all statistical properties of the evolved fluctuation distribution.

In general, the full density contrast is a sum of terms to all orders, $\delta = \delta^{(1)} + \delta^{(2)} + \delta^{(3)} + \dots$, where $\delta^{(n)} \sim D^n(t) \delta_0^n$. In Fourier space the general term can be written as a convolution [4], [9],

$$\tilde{\delta}^{(n)}(k) = \int \frac{d^3 k_1'}{(2\pi)^3} \dots \frac{d^3 k_n'}{(2\pi)^3}[(2\pi)^3 \delta_D(\Sigma k_i' - k)]G_n(k_1', \dots, k_n')\tilde{\delta}(k_1')\dots\tilde{\delta}(k_n'), \qquad (7)$$

where the kernels G_n (ratios of scalar products of the k_i) can be obtained to arbitrary n from a recursion [9], [10] and δ_D is the Dirac δ-function. This has been shown to lead to nonvanishing order to produce the pattern in Eq. (1), with specific values for the amplitude coefficients Q_p and S_p. For spherical average of a distribution with finite $\xi(0)$, the first few are [4], [8] $Q_3 = 34/21$ and $S_4 = 7589/2646$. Coefficients to arbitrary order can be computed from a generating function [11].

At linear order an initially Gaussian distribution remains Gaussian. However, as evidenced by the nonvanishing higher-order correlations generated by nonlinear mode couplings, the evolved probability distribution is no longer Gaussian, but is skewed positively, with much longer large density tails than for the Gaussian distribution. A convenient specific example of a distribution with this correlation pattern is the Gamma distribution, the continuous limit of the discrete negative binomial distribution, with probability density

$$f(x) = \frac{k^k}{\Gamma(k)}x^{k-1}e^{-kx}, \qquad (8)$$

where $x = \rho/\bar{\rho}$ and $\xi = 1/k$. For this distribution the pattern $\xi_p = S_p \xi^{p-1}$ holds exactly to all orders [12], with $S_p = (p-1)!$. FIGURE 1 shows the transition in the behavior of this distribution function for ξ from 1/256 to 4, compared with a Gaussian distribution with the same variance for $\xi < 1$. Another example of a transition from Gaussian to non-Gaussian behavior is given elsewhere in this volume [13].

Perturbation theory also allows the calculation of the local three-point function and its Fourier space counterpart, the bispectrum, which have a characteristic dependence on configuration shape that can be identified as a signature of structure shaped

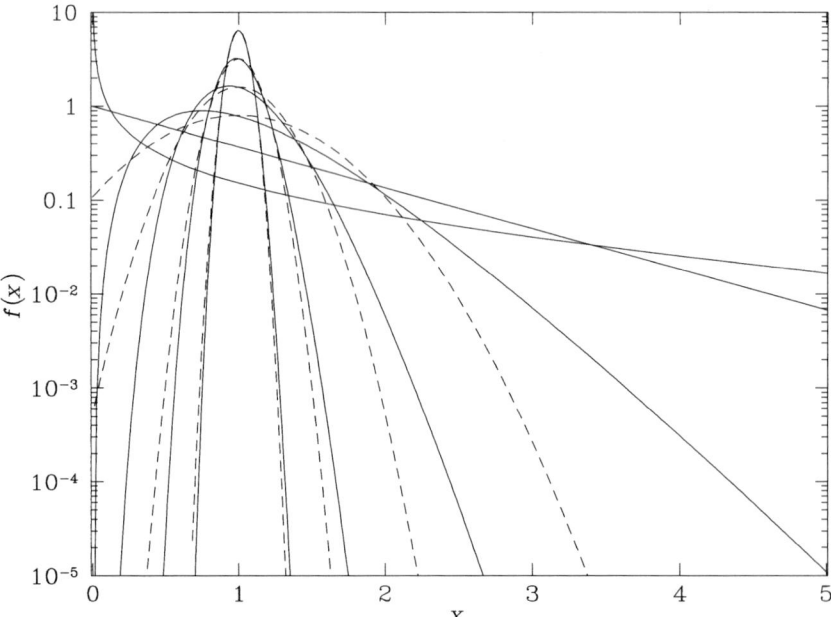

FIGURE 1. Probability density $f(x)$ for the Gamma distribution with variance $\xi = 1/256$, 1/64, 1/16, 1/4, 1, and 4 (narrowest to widest). *Solid lines* show the actual distribution, while for the first four, *dashed lines* show a Gaussian distribution with the same variance. Departures from Gaussian are small for small variance, but grow as ξ increases. For $\xi \geq 1$, there is not even a peak at $\langle x \rangle = 1$.

by gravitational instability [4]. Systematic calculations of p-point amplitudes can be extended to filtered power-law initial spectra with $P(k) \sim k^n$, $\xi(r) \sim r^{-(3+n)}$, for which $S_3 = 34/7 - (3 + n)$, etc. [14]; to universes with $\Omega \neq 1$, for which there are only very weak corrections [15], [16]; and to redshift space, where there are additional peculiar velocity distortions [17]. Yet, even when taken to high order, perturbation theory does not suffice to describe the evolution of structure to arbitrarily large amplitude. This is in part because the theory itself breaks down when the amplitude of the density contrast becomes too large [18], and in part because unless additional terms are added explicitly it does not include shell-crossing [19]. Direct assault on the fully nonlinear regime is the subject of the following section.

3. STRUCTURE IN THE STRONGLY NONLINEAR REGIME

In the regime of strong clustering, exact solutions to the nonlinear equations are notoriously difficult. Nonetheless, considerable understanding can be obtained from a combination of physical reasoning, conceptual models, and a liberal application of symmetries. Even a quick examination of the results of numerical simulations shows that one feature of the evolved distribution is the appearance of tightly bound halo systems that, once formed, maintain a separate identity as more or less independent stable objects, much as the solar system is not much affected by the rest of the galaxy or the universe at large. These are also the systems that have the largest density, and thus plausibly determine the behavior of correlation statistics on the scales they span. Three models based on these considerations are examined.

A. Power-Law Cluster Profiles

An obvious first step is to compute the correlations of matter contained in such halos, and indeed a model with a long history [8], [21], [21] considers mass distributed in randomly placed clusters of specified profile $u(r)$. For clusters with a unique size and density profile, the connected p-point density correlation function is [8], [21], [22]

$$\xi_p(r_1, ..., r_p) = \frac{n \int d^3r \, u(r - r_1) ... u(r - r_p)}{[n \int d^3r \, u(r)]^p}. \qquad (9)$$

Since the observed galaxy two-point correlation function is a power of separation $\xi(r) \sim r^{-\gamma}$, a reasonably guess is that the profile is also a power law, $u(r) \sim r^\varepsilon$. A simple power-counting scaling then gives $\xi_p \sim r^{-\gamma_p}$ for all p on small scales, with power index

$$\gamma_p = p\varepsilon - 3 \qquad (10)$$

for $\varepsilon > 3/2$. This would reproduce $\gamma = 1.8$ if $\varepsilon = 2.4$, but then gives $\gamma_3 = 4.2$ and $\gamma_4 = 6.6$, considerably different than what is found in observations [1]–[3]. A superposition of cluster sizes and masses does not change the nature of this result, also confirmed by replacing halo cores with synthetic power-law profiles in numerical simulations [24]. A recent modification of the halo model is considered in the phenemonological halo model below.

B. Stability, Self-similarity, and Scale Invariance

A second model, built on the concept of stable clustering, leads to results that can be viewed as a continuous clustering hierarchy. Analytically, the desired results follow from particle conservation; physical reasoning and symmetries, including stable clustering, scale invariance, and self-similarity, then are invoked to constrain the form of the solution [8], [25], [26]. In the strong clustering regime, the p-particle equation of continuity is

$$\frac{\partial \xi_p}{\partial t} + \frac{1}{a}\frac{\partial}{\partial x_a^i}(v_a^i \xi_p) = 0 \qquad (11)$$

(implicit sum of repeated indices). Stable clustering implies that gravitationally bound systems are not expanding or contracting with time, so that the peculiar velocity v cancels the Hubble expansion, $v = -\dot{a}x$. After noting that in a homogeneous universe we expect no dependence of ξ_p on the location of the center of mass but only on the $p - 1$ relative positions, this gives

$$a\frac{\partial \xi_p}{\partial a} = 3(p-1)\xi_p + x_a^i \frac{\partial \xi_p}{\partial x_a^i}, \tag{12}$$

relating the rate of growth in time to the rate of change in space. The right-hand side can be evaluated when ξ_p is a scale-free power law, $\xi_p(\lambda x_i) = \lambda^{-\gamma_p} \xi_p(x_i)$, so that

$$\frac{d}{d\lambda}\xi_p(\lambda x_i)\Big|_{\lambda=1} = \sum x_a^i \frac{\partial \xi_p}{\partial x_a^i} = -\gamma_p \xi_p. \tag{13}$$

With $a(\partial \xi_p / \partial a) = \beta_p \xi_p$, this gives the relation

$$\beta_p = 3(p-1) - \gamma_p. \tag{14}$$

Further progress can be made by noting that the full system of evolution equations admit a self-similar solution, such that the general dependence on scale x and scale factor $a(t)$ is in fact a function of the single combination, $\xi(a, x) = \xi(s)$, where $s = x/a^\alpha$, so that $\beta_p/\gamma_p = \alpha$ for all p. This allows us to relate the large scale, perturbation theory result $\xi \sim a^2/x^{(3+n)}$ to the small scale result $\xi \sim a^\beta/x^\gamma$, so that $\beta/\gamma = 2/(3 + n)$, or

$$\gamma = \frac{9 + 3n}{5 + n}. \tag{15}$$

Similar arguments for ξ_p give the general result

$$\gamma_p = (p - 1)\gamma. \tag{16}$$

Equation (16) is an appealing result: the stability of bound systems is intuitively attractive, the observed behavior of correlation functions on small scales is at least approximately as a power law, and the fastest growing mode order by order in perturbation theory is self-similar for the Einstein–de Sitter case with $P(k) \sim k^n$. It implies that the ratio $Q_p = \xi_p/\xi^{p-1}$ is independent of scale, just as was found, for a completely different reason, in the quasilinear regime for nearly Gaussian initial conditions, a result consistent with observations. However, it is not so clear that scale invariance and self-similarity apply to initial conditions with an intrinsic scale, such as the family of cold dark matter models, and to a universe where the cosmological scale factor $a(t)$ is also not a power of time, as $\Omega_m < 1$. Further, studies of perturbations about the self-similar solution have shown that although the full dynamic equations admit such a solution, this solution is not uniquely selected, nor is it necessarily stable [27].

The two descriptions of clustering, as a superposition of power-law clusters and as a self-similar fractal, allow visual representation. FIGURES 2 and 3 show synthetic point distributions for these two models, the power-law cluster and the self-similar fractal.

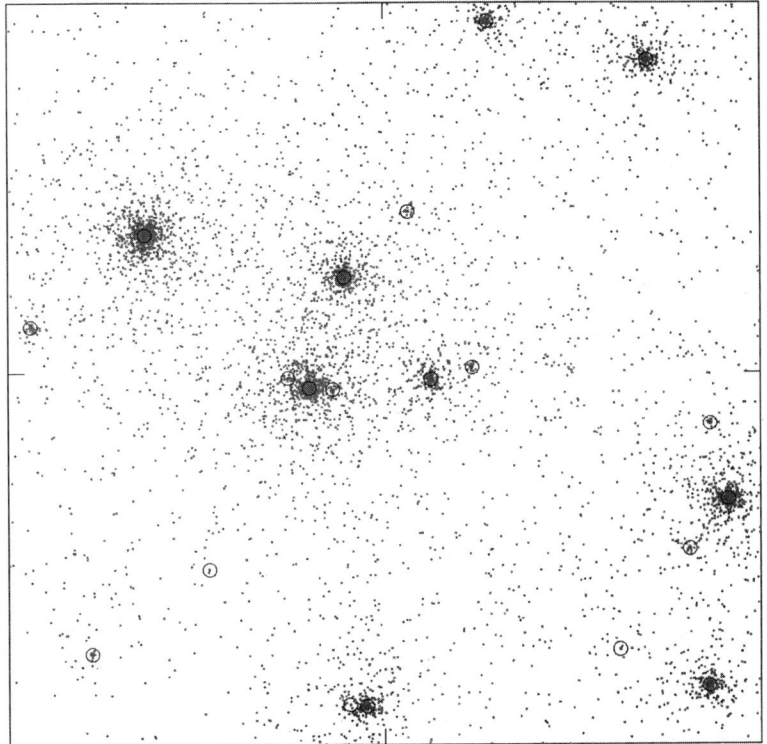

FIGURE 2. A distribution of points in randomly placed clusters with power-law density profile $\rho \sim r^{\varepsilon}$, for $\varepsilon = 2.4$. Cluster centers are marked with circles. Cluster sizes range from 12 to 6500.

C. Phenomenological Halo Model

The models of the previous two sections have achieved some successes but also have shortcomings. Recently it has been realized that there is a phenomenological extension of the halo model that has similar physically motivated origins but at least potentially can also reproduce the predictions of the scale-invariant theory [24], [28]–[32]. The difference between this model and the power-law cluster model above begins with the shape of the halo profile, $\rho/\bar{\rho} = \bar{\delta}u(r/r_s)$, where the amplitude $\bar{\delta}$ and scale r_s are functions of the halo mass, and the profile $u(x)$ is a universal function but not a simple power law [33], [34], $u(x) = 1/x/(1+x)^2$ or $u(x) = x^{-3/2}/(1+x^{3/2})$. The scale r_s is specified by a concentration parameter, $r_s = r_{200}/c$, where r_{200} is the scale for which the average overdensity is a factor of 200. Often c is a power law $c \sim M^{-\beta}$; in particular, $c = (M/M^*)^{-(3+n)/6}$ in scale free models, where M^* is the scale of nonlinearity.

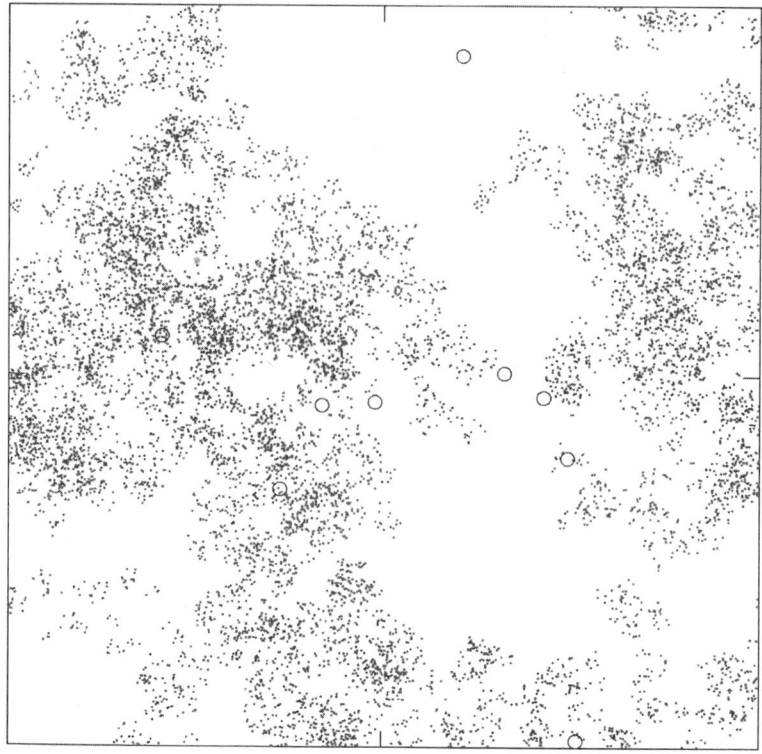

FIGURE 3. A randomly generated self-similar distribution of points with two-point correlation $\xi \sim r^{-\gamma}$, $\gamma = 1.8$.

The extended construction also requires the distribution of the number density of halos with mass M, often approximated by the Press–Schechter [35] prescription,

$$\frac{dn}{dM} = \sqrt{\frac{2}{\pi}} \frac{d\ln\nu}{d\ln M} \frac{\bar{\rho}}{M^2} \nu e^{-\nu^2/2} \qquad (17)$$

in terms of $\nu = \delta_c/\sigma$ for overdensity threshold $\delta_c = 1.68$ and $\sigma^2 = \xi(M)$, or various refinements [36]–[38] that have behavior ν^α instead of ν^1 as $\nu \to 0$. The construction also requires the correlations of halos of masses M_1, M_2; on large scales these are expected to be biased relative to the large scale correlations of density, $\xi_{\text{halo}}(r; M_1, M_2) = b(M_1)b(M_2)\xi_{\text{lin}}(r)$, where $\xi_{\text{lin}}(r)$ is the density correlation function in linear theory and the bias factor $b(M)$ is $b(M) = 1 + (\nu^2 - 1)/\delta_c$ [39], again with possible refinements [40].

The two-point correlation function and power spectrum of mass in the superposition of halos with given profile, mass distribution, and correlation properties contains contributions from particles within a single halo and from two distinct, correlated halos, $\xi(r) = \xi_{1h}(r) + \xi_{2h}(r)$ or in the Fourier domain $P(k) = P_{1h}(k) + P_{2h}(k)$ [28]:

$$\xi(r_{12}) = \int d^3r' \int dM \frac{dn}{dM} \bar{\delta}^2 u(|r'-r_1|/r_s) u(|r'-r_2|/r_s)$$
$$+ \int d^3r' d^3r'' \xi_{\text{lin}}(|r'-r''|) \left[\int dM' \frac{dn}{dM'} \bar{\delta}' u(|r'-r_1|/r_s') b(M') \right] \quad (18)$$
$$\times \left[\int dM'' \frac{dn}{dM''} \bar{\delta}'' u(|r''-r_2|/r_s'') b(M'') \right],$$

$$P(k) = \int dM \frac{dn}{dM} [r_s^3 \bar{\delta}(M) \tilde{u}(kr_s)]^2$$
$$+ \left[\int dM \frac{dn}{dM} r_s^3 \bar{\delta}(M) \tilde{u}(kr_s) b(M) \right]^2 P_{\text{lin}}(k). \quad (19)$$

Similar expressions hold for higher order statistics; the three-point function is the sum of one-, two-, and three-halo terms, $\zeta(r_1, r_2, r_3) = \zeta_{1h} + \zeta_{2h} + \zeta_{3h}$, where the separate terms are [28]

$$\zeta_{1h} = \int d^3r \int dM \frac{dn}{dM} \bar{\delta}^3 u(|r-r_1|/r_s) u(|r-r_2|/r_s) u(|r-r_3|/r_s),$$

$$\zeta_{2h} = \int d^3r\, d^3r'\, \xi_{\text{halo}}(|r-r'|) \int dM \frac{dn}{dM} \bar{\delta}^2 u(|r-r_1|/r_s) u(|r-r_2|/r_s)$$
$$\times \int dM' \frac{dn}{dM'} \bar{\delta}' u(|r'-r_3|/r_s') + [\text{sym}(1,2,3)], \quad (20)$$

$$\zeta_{3h} = \int d^3r\, d^3r'\, d^3r''\, \zeta_{\text{halo}}(r, r'r'') \int dM \frac{dn}{dM} \bar{\delta} u(|r-r_1|/r_s)$$
$$\times \int dM' \frac{dn}{dM'} \bar{\delta}' u(|r'-r_2|/r_s') \int dM'' \frac{dn}{dM''} \bar{\delta}'' u(|r''-r_3|/r_s''),$$

and the bispectrum is similarly $B(k_1, k_2, k_3) = B_{1h} + B_{2h} + B_{3h}$, where

$$B_{1h} = \int dM \frac{dn}{dM} [r_s^3 \bar{\delta}\tilde{u}(k_1 r_s)][r_s^3 \bar{\delta}\tilde{u}(k_2 r_s)][r_s^3 \bar{\delta}\tilde{u}(k_3 r_s)],$$

$$B_{2h} = \int dM \frac{dn}{dM} [r_s'^3 \bar{\delta}\tilde{u}(k_1 r_s')][r_s'^3 \bar{\delta}\tilde{u}(k_2 r_s')]$$
$$\times \int dM' \frac{dn}{dM'} R_s'^3 \bar{\delta}' \tilde{u}(k_3 R_s') P_{\text{halo}}(k_3; M, M') + [\text{sym}(1,2,3)], \quad (21)$$

$$B_{3h} = \int dM \frac{dn}{dM} r_s^3 \bar{\delta}\tilde{u}(k_1 r_s) \int dM' \frac{dn}{dM'} r_s'^3 \bar{\delta}' \tilde{u}(k_2 r_s')$$
$$\times \int dM'' \frac{dn}{dM''} r_s''^3 \bar{\delta}'' \tilde{u}(k_2 r_s'') B_{\text{halo}}(k_1, k_2, k_3; M, M', M'').$$

The halo three-point function ζ_{halo} and bispectrum B_{halo} are related to the corresponding functions for the mass density (see references). The one-halo terms in

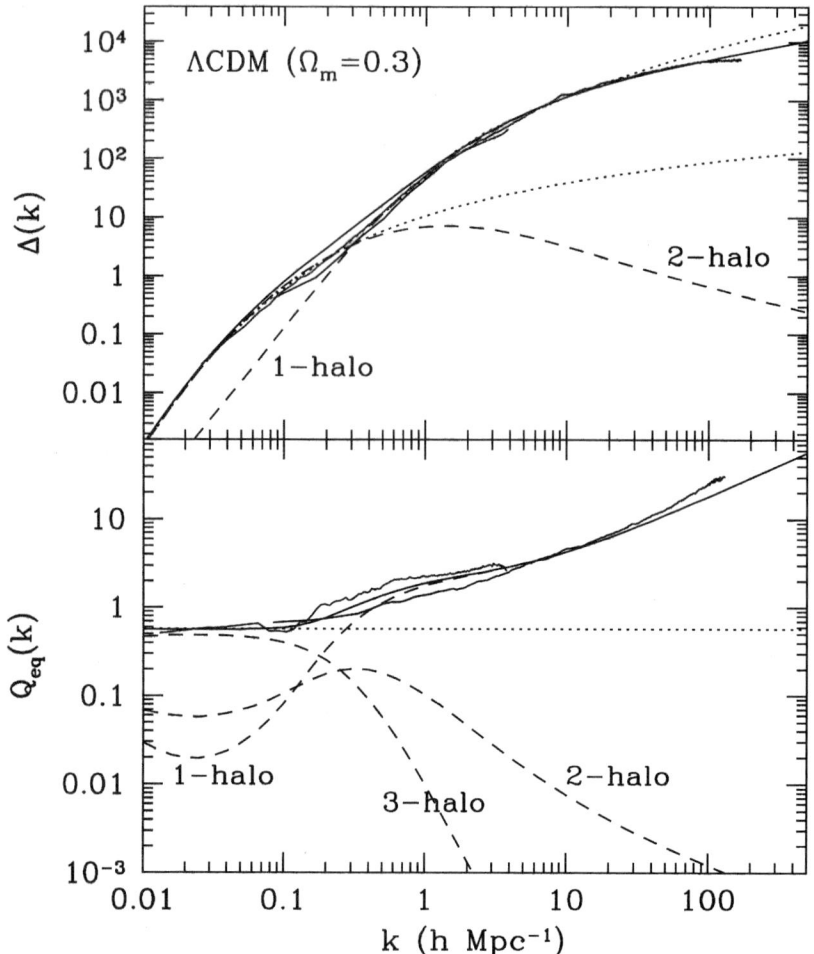

FIGURE 4. N-body results vs. predictions of the analytic model for the power spectrum (*upper*) and bispectrum (*lower*) for low-density CDM simulations with $\Omega_m = 0.3$ and $\Omega_\Lambda = 0.7$ computed from $(100 \text{ Mpc})^3$ and $640 \text{ Mpc})^3$ simulations. The *dashed* curves show the separate contributions to Δ and Q_{eq} computed from the single- and multiple-halo terms of Eqs. (19) and (21); the *solid black* curves show the sum predicted by the model. The *jagged* curves show the N-body results, where synthetic halos have been used to extend the curves to higher k. (The same density profile and $c(M)$ are used for the synthetic halos and the analytic model.)

Eqs. (18) and (19) are generalizations of (9) for a superposition over a distribution halo masses with amplitude and scale a function of mass.

FIGURE 4 shows predictions of the halo model for the power spectrum, plotted as $\Delta(k) = 4\pi k^3 P(k)/(2\pi)^3$, and for the bispectrum for equilateral triangles, $Q_{eq}(k)$, compared with numerical simulations. The halo model successfully describes both two-

and three-point statistics from large scales, in the quasilinear perturbative regime, where these correlations are inherited from the corresponding halo-halo correlations, to small scales, where correlations are dominated by contributions from particles within a single halo. For the specific choices of halo parameters that go into the model in this figure, the large-k behavior of Q_{eq} is not constant, but there is a range of k, spanning approximately the scale of nonlinearity k_{nl} to $c_{nl} k_{nl}$, where $c_{nl} \approx 5$ is the concentration parameter on this scale, where the behavior is determined by the outer halo profile $u \sim x^{-3}$ so that $Q \sim \log k$. The asymptotic, small scale behavior of the halo superposition is intriguingly different from the case of a simple power law profile, and intriguingly similar to that of stable clustering. The difference can be attributed partly to the mass function, dn/dM, and partly to the small-r behavior of the profile, which for the profile shapes considered has $\varepsilon \leq 3/2$. This implies that for large k, the integral over mass for $P(k)$ converges for $\nu \ll 1$, where $dn/dM \sim \nu^\alpha$. The scale r_s depends on mass as $r_s = r_{200}/c \sim M^{1/3}/M^{-\beta} \sim M^{(3\beta+1)/3}$, while (up to possible logarithmic factors) $r_s^3 \tilde{\delta} \sim M$; then $\Delta(k) = 4\pi k^3 P(k)/(2\pi)^3$ goes as $\Delta(k) \sim k^3 \int dM \nu^\alpha \tilde{\delta}^2 (kr_s)$. Changing variables to $y = kr_s$, we see [29] that $\Delta(k)$ scales as k^γ, with

$$\gamma = \frac{18\beta - (3+n)\alpha}{2(3\beta+1)} = \left(\frac{9+3n}{5+n}\right) - \alpha\left(\frac{3+n}{5+n}\right) \quad (22)$$

where $\sigma \sim M^{(3+n)/6}$ and the last equality holds for scale free models with $\beta = (3+n)/6$. This result is independent of the details of the profile shape, provided that the integral is dominated by scales $kr_s \sim 1$. The scaling behavior of the higher-order functions gives

$$\gamma_p = \frac{18(p-1)\beta - (3+n)\alpha}{2(3\beta+1)} = (p-1)\left(\frac{9+3n}{5+n}\right) - \alpha\left(\frac{3+n}{5+n}\right) \quad (23)$$

These agree with the self-similar, stable clustering result for $\alpha = 0$. For $\varepsilon > 3/2$, convergence of the integral over k requires the exponential factor, and is then dominated by scales where $\nu \approx 1$; this reproduces the power law cluster result above. FIGURE 5 shows that combinations of α and β that lead to similar $\Delta(k)$ can be distinguished by the three-point function.

From the behavior of ξ in the pair conservation equation, it can be shown that for $\xi \sim a^2/x^{(3+n)}$ on large scales, the pair velocity at small separation goes to [28]

$$-\frac{v}{Hr} \to \frac{2}{n+3} \frac{18\beta - \alpha(n+3)}{6 + \alpha(n+3)}. \quad (24)$$

This is a constant, but not necessarily 1, the value for stable clustering. It can easily be checked that $-v/Hr$ approaches unity for any pair (α, β) that reproduces the stable clustering result (15) for the two-point function, but only for $\alpha = 0$ does the halo model reproduce (16) for all orders. FIGURE 6 shows the behavior of the pair velocity for various parameter combinations.

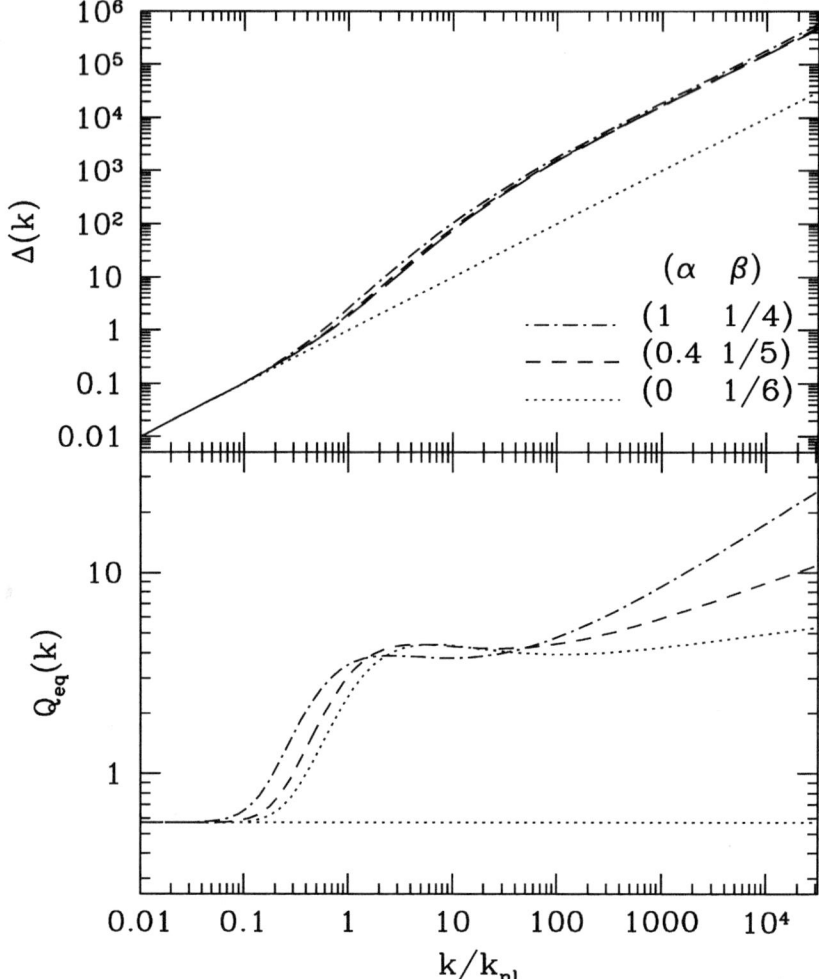

FIGURE 5. The power spectrum (*upper panel*) and three-point amplitude (*lower panel*) given by the analytic halo model for the $n = -2$ scale-free model. Three sets of (α, β) are shown, where α and β parameterize the halo mass function and halo concentration. *Dotted* curves show the lowest-order perturbative predictions. While different combinations of (α, β) give nearly identical $\Delta(k)$ which all agree with the two-point stable clustering result, the three-point amplitude $Q(k)$ has distinct high-k behavior; stable clustering holds to all orders only if $\alpha = 0$ and $\beta = (3 + n)/6$.

4. DISCUSSION

In this paper we have examined the behavior of nonlinear cosmological density fluctuations, ranging from the nearly homogeneous large-scale universe, with small, nearly Gaussian fluctuations in mass density, to a universe that is strongly clustered

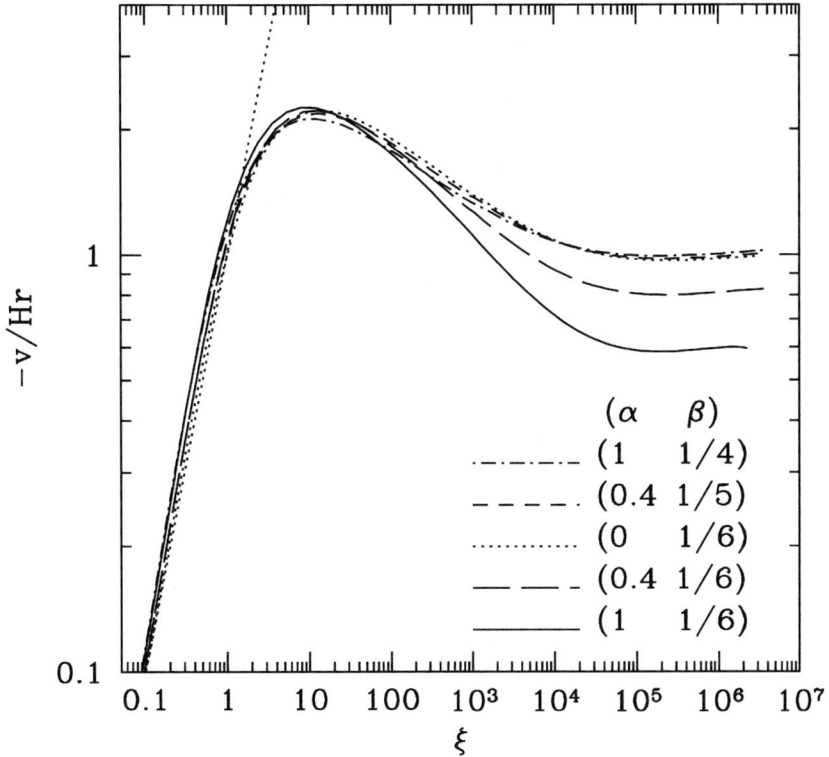

FIGURE 6. Prediction of the analytic halo model for the ratio of pairwise peculiar velocity and Hubble flow, $-v/Hr$, versus the two-point correlation function ξ for a scale free model with $n = -2$. The velocity ratio approaches a constant in the deeply nonlinear regime, but the value is not always unity, as would be required by stable clustering. For example, $-v/Hr \to 1$ at $\xi \gg 1$ for $(\alpha, \beta) = (1, 1/4)$, $(0.4, 1/5)$, and $(0, 1/6)$, but reaches a smaller value for $\alpha = 0.4$ or 1 and $\beta = 1/6$, the range of values found in simulations. The *dotted straight line* shows the linear theory prediction, which is followed accurately at $\xi < 1$.

on small scales. Perturbation theory works well on large scales, and techniques now exist to obtain results to high order. Nonlinear mode couplings generate higher-order correlations of all orders, which are unimportant as $\xi \to 0$ but for $\xi \sim 1$ lead to a fluctuation distribution that is strongly non-Gaussian. As parametrized by the variance ξ, the transition from one behavior to the other is continuous, as modeled in FIGURE 1.

On small scales, where perturbation theory breaks down, it is intuitively appealing that on these scales the nature of the distribution is determined by the tightly bound halos. The power law cluster model is a first attempt to realize structure as a superposition of such halos, while the self-similar model is built on the concept of stable clustering. Despite their very different visual impressions, both the power-law model and the scale invariant model can be viewed as special cases of a general halo model. The general model contains mass distributed in spherical halos with specified halo–halo correlations, mass function, and radial profile with amplitude and scale functions of the mass derived from numerical simulations. The model may be easily extended to include additional effects, such as substructure and nonspherical halos. The halo model is for the most part specified by quantities defined in the perturbative regime; popular forms of the mass function dn/dM are functions of the linear overdensity $\nu = \delta_c/\sigma$, and halo–halo correlations are significant only on scales where quasilinear theory applies. The strongly nonlinear behavior arises from the single halo term, where the specific profile shape is the one ingredient not known from perturbation theory but determined from numerical simulations; but results for low order correlation statistics do not depend in detail on profile shape, as evident in Eqs. (23) and (24), which do not depend on ε. This model can reproduce the constant correlation amplitudes Q_p in the quasilinear regime, where it inherits the results of quasilinear theory, and in the stable clustering model for some range of parameter values, and for many reasonable values does so at least approximately over a range of scales entering the nonlinear regime. Thus, it may deserve to be characterized as a model that captures the essence of the transition to nonlinearity.

ACKNOWLEDGMENTS

C.-P.M. acknowledges support of an Alfred P. Sloan Foundation Fellowship, a Cottrell Scholars Award from the Research Corporation, a Penn Research Foundation Award, and NSF grant AST 99-73461.

REFERENCES

1. PEEBLES, P.J.E. & E.J. GROTH. 1975. Statistical analysis of catalogs of extragalactic objects. V. Three-point correlation function for the galaxy distribution in the Zwicky catalog Astrophys. J. **196:** 1–11.
2. GROTH, E.J. & P.J.E. PEEBLES. 1977. Statistical analysis of catalogs of extragalactic objects. VII. Two- and three-point correlation functions for the high-resolution Shane-Wirtanen catalog of galaxies. Astrophys. J. **217:** 385–405.
3. FRY, J.N. & P.J.E. PEEBLES. 1978. Statistical analysis of catalogs of extragalactic objects. IX. The four-point galaxy correlation function. Astrophys. J. **221:** 19–33.
4. FRY, J.N. 1984. The galaxy correlation hierarchy in perturbation theory. Astrophys. J. **279:** 499–510.
5. MEIKSIN, A., I. SZAPUDI & A. SZALAY. 1992. Higher order correlations of IRAS galaxies. Astrophys. J. **394:** 87–90.
6. GAZTAÑAGA, E. 1994. High-order galaxy correlation functions in the APM galaxy survey. Mon. Not. R. Astron. Soc. **268:** 913–924.
7. SZAPUDI, I., G.B. DALTON, G. EFSTATHIOU & A. SZALAY. 1995. Higher order statistics from the APM galaxy survey. Astrophys. J. **444:** 520–531.
8. PEEBLES, P.J.E. 1980, The Large Scale Structure of the Universe. Princeton Univ. Press. Princeton, NJ.
9. GOROFF, M.H., B. GRINSTEIN, S.-J. REY & M. WISE. 1986. Coupling of modes of cosmological mass density fluctuations. Astrophys. J. **311:** 6–14.
10. JAIN, B. & E. BERTSCHINGER. 1994. Second-order power spectrum and nonlinear evolution at high redshift. Astrophys. J. **431:** 495–505.
11. BERNARDEAU, F. 1992. The gravity-induced quasi-Gaussian correlation hierarchy Astrophys. J. **392:** 1–14.

12. FRY, J.N. 1985. Cosmological density fluctuations and large-scale structure: From N-point correlation functions to the probability distribution. Astrophys. J. **289**: 10–17..
13. MUTTALIB, K.A. & P.WÖLFLE. 2001. Emergence of anomalous distributions in disordered systems. Ann. N.Y. Acad. Sci. **927**: 136–142.
14. BERNARDEAU, F. 1994. Skewness and kurtosis in large-scale cosmic fields. Astrophys. J. **433**: 1–18.
15. BOUCHET, F.R., R. JUSZKIEWICZ, S. COLOMBI & R. PELLAT. 1992, Weakly nonlinear gravitational instability for arbitrary Omega. Astrophys. J. **394**: L5–L8.
16. BOUCHET, F.R., S. COLOMBI, E. HIVON & R. JUSZKIEWICZ. 1995. Perturbative Lagrangian approach to gravitational instability. Astron. Astrophys. **296**: 575–608.
17. HIVON, E., F.R. BOUCHET, S. COLOMBI & R. JUSZKIEWICZ. 1995. Redshift distortions of clustering: a Lagrangian approach. Astron. Astrophys. **298**: 643–660.
18. FRY, J.N. 1994. The minimal power spectrum: Higher order contributions, Astrophys. J. **421**: 21–26.
19. SCOCCIMARRO, R. 2001. A new angle on gravitational clustering. Ann. N.Y. Acad. Sci. **927**: 13–23.
20. NEYMAN, J. & E.L. SCOTT. 1952. A theory of the spatial distribution of galaxies. Astrophys. J. **116**: 144–163.
21. PEEBLES, P.J.E. 1974. A model for continuous clustering in the large-scale distribution of matter. Astrophys. Space Sci. **31**: 403–410.
22. MCCLELLAND, J & J. SILK. 1977. The correlation function for density perturbations in an expanding universe. II. Nonlinear theory Astrophys. J. **217**: 331–352.
23. SHETH, R.K. & B. JAIN. 1997. The non-linear correlation function and density profiles of virialized haloes. Mon. Not. R. Astron. Soc. **285**: 231–238.
24. MA, C.-P. & J.N. FRY. 2000. Halo profiles and the nonlinear two- and three-point correlation functions of cosmological mass density. Astrophys. J. **531**: L87–L90.
25. PEEBLES, P.J.E. 1974. The gravitational-instability picture and the nature of the distribution of galaxies. Astrophys. J. **189**: L51–L55.
26. DAVIS, M. & P.J.E. PEEBLES. 1977. On the integration of the BBGKY equations for the development of strongly nonlinear clustering in an expanding universe. Astrophys. J. Suppl. Ser. **34**: 425–450.
27. RUAMSUWAN, L. & J.N. FRY. 1992. Stability of scale-invariant nonlinear gravitational clustering. Astrophys. J. **396**: 416–429.
28. MA, C.-P. & J.N. FRY. 2000. What does it take to stabilize gravitational clustering? Astrophys. J. **538**: L107–L111.
29. MA, C.-P. & J.N. FRY. 2000. Deriving the nonlinear cosmological two- and three-point correlation functions from analytic dark matter halo profiles and mass functions. Astrophys. J. **543**: 503–513.
30. SELJAK, U. 2000. Analytic model for galaxy and dark matter clustering. Mon. Not. R. Astron. Soc. **318**: 203–213.
31. PEACOCK, J.A. & R.E. SMITH. 2000. Halo occupation numbers and galaxy bias. Mon. Not. R. Astron. Soc. **318**: 1144–1156.
32. SCOCCIMARRO, R., R.K. SHETH, L. HUI & B. JAIN. 2001. How many galaxies fit in a halo? constraints on galaxy formation efficiency from spatial clustering. Astrophys. J. **546**: 20–34. 1
33. NAVARRO, J.F., C.S. FRENK & S.D.M. WHITE. 1997. A universal density profile from hierarchical clustering. Astrophys. J. **490**: 493–508.
34. MOORE, B., F. GOVERNATO, T. QUINN, J. STADEL & G. LAKE. 1999. Resolving the structure of cold dark matter halos. Astrophys. J. **499**: L5–L8.
35. PRESS, W.H. & P. SCHECHTER. 1974. Formation of galaxies and clusters of galaxies by self-similar gravitational condensation. Astrophys. J. **187**: 425–438
36. SHETH, R. K. & G. TORMEN. 1999. Large-scale bias and the peak background split. Mon. Not. R. Astron. Soc. **308**: 119–126.
37. SHETH, R. 2001. A random walk through models of nonlinear clustering. Ann. N.Y. Acad. Sci. **927**: 1–12.
38. SHANDARIN, S. F. 2001. The cosmological mass function in the Zel'dovich approximation. Ann. N.Y. Acad. Sci. **927**: 70–83.

39. MO, H.J., Y.P. JING & S.D.M. WHITE. 1996. The correlation function of clusters of galaxies and the amplitude of mass fluctuations in the Universe. Mon. Not. R. Astron. Soc. **282:** 1096–1104.
40. JING, Y.P. 1998. Accurate fitting formula for the two-point correlation function of dark matter halos. Astrophys. J. **503:** L9–L12.

Index of Contributors

Brisudova, M., 127–135

Feldman, H.A., 43–53
Fry, J.N., 143–158

Gaztañaga, E., 24–42, 110–126
Gorman, P., 43–53

Juszkiewicz, R., 24–42

Kinney, W.H. 102–109, 127–135

Lobo, J.A., 110–126

Ma, C.-P., 143–158
Melott, A.L., 43–53
Muttalib, K.A., 136–142

Scoccimarro, R., 13–23
Shandarin, S.F., 70–83
Sheth, R.K., 1–12
Sikivie, P., 102–109
Szapudi, I., 94–101

Verde, L., 54–69

Watkins, R., 43–53
Wölfle, P., 136–142

Zaldarriaga, M., 84–93

OHIO UNIVERSITY LIBRARY
book as